Jose Araujo Blanco

The Petri dish

Jose Araujo Blanco

The Petri dish

A Journey Through Microbiological Time and Space

ScienciaScripts

Imprint
Any brand names and product names mentioned in this book are subject to trademark, brand or patent protection and are trademarks or registered trademarks of their respective holders. The use of brand names, product names, common names, trade names, product descriptions etc. even without a particular marking in this work is in no way to be construed to mean that such names may be regarded as unrestricted in respect of trademark and brand protection legislation and could thus be used by anyone.

Cover image: www.ingimage.com

This book is a translation from the original published under ISBN 978-613-9-41110-8.

Publisher:
Sciencia Scripts
is a trademark of
Dodo Books Indian Ocean Ltd. and OmniScriptum S.R.L publishing group

120 High Road, East Finchley, London, N2 9ED, United Kingdom
Str. Armeneasca 28/1, office 1, Chisinau MD-2012, Republic of Moldova, Europe
Printed at: see last page
ISBN: 978-620-7-92906-1

PROFESSOR JOSÉ AMABLE ARAUJO

Biologist graduated from the Universidad del Zulia with a Master's degree in Microbiology, Lawyer graduated from the Universidad Bolivariana de Venezuela, with a Master's degree in Museology, and Violinist. He has made stays in several countries, he was professor winner of competition of the Universidad Francisco de Miranda in the area of microbiology dictating the chair of microbiology, for Environmental Sciences, Restoration of Cultural Furniture, Nursing, Veterinary Medicine and Medicine, was Head of the Research Unit in Electron Microscopy for 8 years attached to the Dean of Research, coordinator of the area of microbiology for 10 years, He is currently a professor at the University of Magdalena, where he teaches violin and is the director of the string group and the symphony of the University of Magdalena. He also teaches Microbiology for Biology students in the area of Health Sciences, is the main professor of Microbiology in Dentistry, teaches Biology and Biochemistry, and is also an associate professor of Microbiology for the Medicine programme. He has a career spanning over 18 years as a lecturer in Microbiology and its associated areas.

TABLE OF CONTENTS

DEDICATION 4

CHAPTER 1 5

CHAPTER 2 43

CHAPTER 3 70

BIBLIOGRAPHICAL REFERENCES 100

DEDICATION

I dedicate this book to my parents, who have always been there for me, without hesitation, so that I could effectively acquire the academic, technical, scientific and artistic knowledge and skills that allow me to make my way in life. To my wife Yarubit Teresa Rojas Montilla, inseparable companion of so many dreams, to my daughters Dr. Maria de los Angeles, Maria Gabriela and Maria Auxiliadora, I love them, also with special appreciation I dedicate this book.
To my dear Frank.......

Dr. José Francisco Yegres is a great microbiologist, mycologist and scientist, especially the best person from whom I have learned and continue to learn.
To all of them I dedicate this book of interest that can surely fill a gap on an exciting subject and that I try to relate in a more intimate and digestible language for general readers interested in microbiological science.

CHAPTER 1

THE ANCIENT

Microscopic Roots: Petri dish traversal along the Time The microbial universe, that space that cannot be seen with the naked eye, is undoubtedly a place where it is possible to discover beings that on a microscopic scale reign with influence even in our lives. Knowing this vast world is undoubtedly an experience that can only be explained by those who, from experience, get to know the microbial world. In this context we will deal with the Petri dish, an invention with more than 100 years of validity that marked the beginning, development and possible future of microbiological science. From its introduction into the scientific arena to its current impact on our understanding of microbiology, the Petri dish has been a fundamental tool in the study of microorganisms. Initially, its use in microbiology revolutionised the observation and cultivation of microorganisms, and thanks to pioneers such as Julius Richard Petri and Robert Koch, it laid the indispensable technical foundations that are used in laboratories all over the world.

Over time, the Petri dish culture technique has evolved from rudimentary forms to advanced methodologies. used today. This development has improved the efficiency and accuracy of microbiological research, transforming our understanding of microbiology and allowing us to explore microorganisms that have existed for millennia. In addition, historical case studies and key discoveries have demonstrated the significant impact of the Petri dish on the advancement of microbiology. These examples illustrate its crucial role in the development of scientific knowledge and highlight its importance in the understanding and control of disease, the production of food and medicines, and the preservation of the environment. This first chapter will delve into the rich history and innovative potential of the Petri dish, preparing us to explore future advances and challenges in modern microbiology.

1.1 Introduction to the historical use of the Petri Dish

Imagine a small transparent container that acts as a miniature world for micro-organisms: the Petri dish. This ingenious 19th century invention has been like a window into the invisible realm of bacteria, fungi and other tiny creatures that populate our planet.

"Eine kleine Modification des Koch'schen Plattenverfahrens", 1887. "A small modification of Koch's plate method", 1887 was the title of the publication that changed the way microbiology is viewed and which is still in use today and possibly in the future. Think of the Petri dish as a home for these micro-organisms, a place where they can grow and multiply under controlled conditions. It is like a city in which each micro-organism finds a space as if it were its own neighbourhood in the culture medium that fills the dish. I have always thought of the Petri dish as a small universe or a small world with actors playing a role that we study in order to understand the microbial basis and use it to the benefit of the humanity, in fact, is seen as a space for sociological analysis and social interaction by some sociologists. But why is this little plate so important? Well, it turns out that it has helped us understand and fight disease, produce food more safely and protect our environment.

Think you have a sample of water and you want to know if it contains bacteria that could make you sick. You simply take some of that water and spread it over the Petri dish. Then you let the bacteria in the water multiply and form visible colonies. Voilà! Now you can see if the water is contaminated and take steps to purify it. If you want to know more, here is a list of Current Culture Media with which you can test your water:

Pepton Water 0.1%: A liquid culture medium commonly used for the recovery of microorganisms present in water samples. Its simple composition makes it suitable for the recovery of a wide variety of microorganisms, including enterobacteria. Brilliant Green Bile Lactosate Broth (VRBL): This liquid medium is used for the detection of total coliforms and faecal coliforms in water samples. The presence of coliforms results in a change of colour of the medium, which facilitates their identification. Methylene Blue Eosin Lactosate Broth (MLEB): Other liquid medium used

for the detection of total coliforms and faecal coliforms. Like the VRBL, the change in colour of the medium indicates the presence of coliforms. Tergitol 7 Lactate Broth (TBX): This liquid medium is used for the recovery of coliform bacteria and other micro-organisms indicative of contamination. faecal matter in water samples. Green Vibrant Bile Broth (GVBB): A liquid medium used for the detection and

confirmation of faecal coliforms in water samples. EC Broth (Escherichia coli): This selective liquid medium is used for the detection and confirmation of E. coli in water samples.

Eosin Methylene Blue Agar (EMB): A solid medium used for the detection and enumeration of faecal coliforms, including E. coli, in water samples. The E. coli colonies appear a characteristic dark green metallic colour. Chromocult Agar (CCA): A solid medium used for detection and differentiation. coliforms and E. coli in water samples. Its ability to distinguish between different groups of bacteria makes it useful in microbiological water analysis. Salmonella-Shigella Agar (SSA): A solid, selective medium used for the detection of Salmonella and Shigella in water and other food samples. Hektoen Enteric (HE) Agar: A selective solid medium used for the detection and differentiation of pathogenic enteric bacteria, such as Salmonella and Shigella, on water and food samples. Lauryl Sulfate Broth (LSB) Agar: This liquid medium is used for the detection of total coliforms and faecal coliforms in water samples. The presence of coliforms produces turbidity in the medium, indicating possible contamination.But the Petri dish not only helps us to detect micro-organisms, it also allows us to study them closely. We can observe how they behave, what makes them grow and what kills them.   This has allowed us to allowed the development of drugs and vaccines to combat diseases caused by bacteria and other pathogens.In addition, the Petri dish has been an invaluable tool in the food industry. Imagine for a moment that you work in a dairy factory and you want to make sure that your products are free of harmful bacteria. With the help of the Petri dish, you can monitor the quality of your products and ensure their safety.

The Petri dish is much more than just a piece of plastic or glass. It is a powerful tool that has allowed us to explore a world invisible to the naked eye and has helped us protect public health, produce safe food and preserve our environment. So next time you see a Petri dish in a lab, remember what it represents and the crucial role it plays in science and in our lives.

1.2 Early applications in microbiology and bacteriology

In 1887, when Julius Richard Petri created his Petri dish, the world was experiencing a period of major scientific and technological advances. The second half of the 19th century witnessed a series of revolutionary discoveries in various areas of science, including biology and medicine.

At the time, the microbial theory of disease proposed by Louis Pasteur and Robert Koch was gaining recognition. Pasteur had demonstrated the connection between microorganisms and fermentation, while Koch had identified specific microorganisms as causing specific diseases, thus laying the foundations of medical bacteriology.

The development of new laboratory techniques was also booming in the 1880s. Koch had introduced agar as a solidifying agent in culture media, which allowed the growth of bacterial colonies on a solid medium, facilitating their observation and study. In this historical context, Julius Richard Petri created his Petri dish as an innovative tool for the cultivation of micro-organisms in a controlled, sterile environment. His invention greatly simplified the bacterial culture process, allowing scientists to isolate and study micro-organisms with greater ease and precision.

Therefore, the creation of the Petri dish in 1887 represented an important milestone in the history of microbiology and medicine, as it provided an invaluable tool for the study of microorganisms and their role in human and animal health.

It is thus a time of departure in the history of microbiology, a time when human curiosity encountered the invisible world of microorganisms. On this journey, we find ourselves at the dawn of bacteriological science, an unknown territory where every discovery was a step towards understanding life in its tiniest and most mysterious form. Early applications in microbiology and bacteriology take us back to a time when pioneering scientists explored the mysteries of the microscopic with ingenuity and determination. These applications, though simple in comparison to modern technologies, laid the foundation for our current understanding of microbial life and its impact on the world around us.

Thus the first applications of the Petri dish emerge as a novel tool and an instrument that allowed scientists to observe and study microorganisms in a systematic and controlled way. With its simple but ingenious design, the Petri dish opened the door to a new world of possibilities in microbiological research.

In fact, the origin of the Petri dish is linked to the collaboration between Julius Richard Petri and his mentor, the eminent microbiologist Robert Koch. In the effervescent atmosphere of 19th century microbiology, the Petri innovation received Koch's support and expertise, which contributed to its rapid adoption and worldwide recognition. Today, the Petri dish is a standard item in every microbiology laboratory and is recognised as an indispensable tool. Our imagination may think of the pioneers of microbiology, such as Louis Pasteur and Robert Koch, using the Petri dish to isolate and culture microorganisms in their laboratories. With each culture in a dish, windows were opened to the understanding of disease, the fermentation of food and the decomposition of organic matter. The term "Petri dish" has taken root as a common name in the scientific field, although it is worth noting that it is suggested to write it with an initial capital letter, in recognition of the surname of its inventor. Furthermore, it is important to point out the improper and decontextualised use of this term, underlining its correct application in the proper context. Julius Richard Petri, born on 31 May 1852 in Barmen, played a crucial role in the development of this fundamental tool. He studied medicine in Berlin and worked as a laboratory assistant to Robert Koch, where he was able to apply and refine his innovative ideas.

The Petri dish, in essence, is a modification of the method introduced by Koch, which used gelatin to solidify liquid culture media. Koch's technique, presented at the Seventh International Medical Congress in London in 1881, received widespread praise and contributed significantly to its global adoption, even praised by Pasteur himself. Petri, building on this method, introduced flat double dishes with a slightly larger lid, which facilitated the handling and observation of microbial cultures.

From the identification of pathogens to the study of bacterial resistance, early applications of the Petri dish laid the groundwork for future advances in medicine, agriculture and the food industry. These small glass tools became mainstays of microbiological research, allowing scientists to explore an invisible universe that influences our lives in ways we are only beginning to understand.

Early applications in microbiology and bacteriology, facilitated by the Petri dish, ushered in an era of discovery and understanding in the microbial world. From its use in the identification

From pathogens to their role in basic research, these applications laid the foundation for our current understanding of microbial life and its impact on our world.

The Petri dish has left an indelible mark on the history of microbiology, standing out as an essential tool for more than a century of scientific research. Its design has allowed the cultivation and study of microorganisms, driving significant advances in various branches of science.

Since its invention, the Petri dish has been an indispensable tool in the scientist's arsenal, opening doors to a microscopic world full of secrets and challenges. Its impact goes beyond mere scientific curiosity; it has been the catalyst for monumental advances in the field of microbiology and bacteriology, shaping our understanding of health and disease at a fundamental level.

Just imagine for a moment the world before the discovery of the tuberculosis bacillus by Robert Koch in 1882. Infectious diseases were rampant, often unidentified and misunderstood. The Petri dish possibly allowed Koch to isolate and culture this pathogen, shedding light on its nature and opening the way to effective treatments. This discovery was a cornerstone in the history of medicine, illustrating the transformative power of a simple glass tool. An even more impressive example is the development of vaccines, such as the pioneering smallpox vaccine created by Edward Jenner in the 18th century. Jenner, armed with the Petri dish and his ingenuity, observed that people exposed to the cowpox virus did not contract human smallpox. This discovery, backed by the ability to grow and study microorganisms in Petri dishes, laid the foundation for modern immunisation, saving countless lives throughout history.

But the Petri dish has not only been a weapon in the fight against infectious diseases; it has also been an invaluable tool in the understanding and control of epidemics. In 1854, John Snow used the observation of microorganisms in Petri dishes to identify Vibrio cholerae as the causative agent of cholera, revolutionising our understanding of how diseases spread and laying the groundwork for effective control and prevention measures.

Alexander Fleming's discovery of penicillin in 1928 is another milestone that illustrates the potential of the Petri dish in the development of antibiotics. By casually observing the growth of mould in a dish, Fleming noticed that this mould had antibacterial properties. This finding, made possible by detailed observation of micro-organisms in Petri dishes, paved the way for a revolution in the treatment of infectious diseases and increased life expectancy worldwide. And so we could go on, exploring how the Petri dish has been instrumental in epidemiological research, the study of fermentation, bacterial genetics, the development of staining techniques and the understanding of bacterial resistance to antibiotics. Its role in environmental microbiology research is also crucial, allowing the identification and characterisation of microorganisms present in soil and water samples, and shedding light on the complex microbial ecosystems around us.

Ancient research developed from the invention of the Petri dish:

Application	Description
Discovery of pathogenic micro-organisms	The Petri dish has been fundamental in the identification of disease-causing micro-organisms, such as the discovery of the tuberculosis bacillus by Robert Koch in 1882.
Vaccine development	The ability to grow micro-organisms in Petri dishes has facilitated the development of vaccines, such as the smallpox vaccine developed by Edward Jenner in the 18th century.
Control of infectious diseases	The study of microorganisms in petri dishes has been crucial for the control of infectious diseases, such as the identification of Vibrio cholerae as the causative agent of cholera. in 1854 by John Snow.
Antibiotic development	The observation of micro-organisms in Petri dishes has led to the discovery of antibiotics, such as penicillin by Alexander Fleming in 1928.
Research in epidemiology	The ability to grow bacterial cultures in Petri dishes has made it possible to study the epidemiology of diseases
	infectious diseases, such as the cholera outbreak in London in 1854. researched by John Snow.

Study of fermentation	Petri dishes have been used to study fermentation, a fundamental process in the production of food and beverages, such as lactic acid fermentation in the yoghurt production.
Bacterial genetics research	The observation of bacteria in Petri dishes has been crucial to the study of bacterial genetics, such as the gene transfer between bacteria studied by Joshua Lederberg in the1940s.
Development of colouring techniques	The Petri dish has been used in the development of techniques for bacterial staining, such as the Gram stain developed by Hans Christian Gram in 1884.
Study of bacterial resistance	Petri dishes have been used to study bacterial resistance to antibiotics, such as the observation of strains resistant Staphylococcus aureus in the 1940s.
Research in environmental microbiology	The ability to grow micro-organisms in Petri dishes has been fundamental to the study of environmental microbiology, as the identification

1.3 Pioneering contributions by Julius Richard Petri and Robert Koch

The Petri dish is much more than just a glass container, it is a window into the invisible world of micro-organisms and a tool that has radically transformed our understanding of health and disease. Its legacy will live on for generations, reminding us of the power of meticulous observation, scientific curiosity and perseverance in the pursuit of knowledge. Julius Richard Petri's original publication on the Petri dish, entitled "Eine kleine Modification des Koch'schen Plattenverfahrens" ("A small modification of the Koch's plate method"), is set in a particular historical and scientific context. In 1887, when Petri published his article, Germany was experiencing rapid industrial growth and expansion in science. The nation was in a period of significant change, both politically and culturally. Under the leadership of Otto von Bismarck, the German Empire had only recently been established (in 1871) and was emerging as an industrial and military power in Europe.

In 1887, when Julius Richard Petri published his article on the Petri dish, Germany was under the leadership of Kaiser Wilhelm I as Emperor and Otto von Bismarck as Chancellor. Wilhelm I was the first Emperor of the German Empire and his reign lasted until his death in March 1888. After his death, his son, Frederick III, assumed the throne, but his reign was very brief, as he died in June of the same year. Subsequently, Wilhelm II, the son of Frederick III, became emperor, continuing with Otto von Bismarck as his chancellor for a short period until 1890, when Bismarck was deposed.

The political climate in Germany at the time was one of consolidation and strengthening of the new empire, with an emphasis on industrial and scientific progress. Otto von Bismarck's policy, known as "Realpolitik", promoted an environment of stability and pragmatism, which allowed for remarkable development in various scientific fields. This environment favoured innovations, as state support and investment in scientific and educational infrastructure boosted the advancement of microbiology and other sciences. Political stability and Bismarck's vision for a strong, modern state contributed to Germany's emergence as a leading centre of scientific research during this period.

At the time, microbiology was in full swing. Robert Koch, Petri's mentor, had gained renown for his discoveries about infectious diseases and his innovations in laboratory. Koch's publication in 1882 of the discovery of the tubercle bacillus had generated great interest in the study of micro-organisms and their roles in health and disease.

The journal in which Petri published his article, the "Centralblatt für Bakteriologie und Parasitenkunde" (Central Journal of Bacteriology and Parasitology), was one of the leading scientific journals of the time, dealing with relevant topics in microbiology and parasitology. The inclusion of Petri's article in this journal indicated its importance and relevance within the scientific community of the time.

Petri's contribution with his invention of the Petri dish represented a significant advance in microbiology by providing a simple and effective tool for the cultivation and study of microorganisms. His publication in 1887 marked the beginning of a new era in microbiological research and laid the foundation for future advances in the field.

This collaboration between Petri and Koch was crucial for the advancement of this young science - microbiology. Together, they explored new cultivation techniques that would forever change the way microorganisms are studied. Recreate in your mind's eye Koch's laboratory at the Kaiserliches Gesundheitsamt in 19th century Germany: a vibrant environment of discovery where brilliant minds immersed themselves in the search for innovative solutions to the scientific challenges of the time.

Petri devised the Petri dish in 1887 while working in Koch's laboratory as an innovative solution for growing bacteria efficiently. Its simple but effective design consisted of a shallow container with a lid that allowed microorganisms to grow in a solid culture medium. This invention greatly facilitated the study of microorganisms by providing a controlled environment for their growth and observation.

In the midst of this atmosphere of scientific ferment, Petri had the inspiration for his revolutionary invention. Realising the limitations of existing culture methods, he devised an elegant and practical solution: the Petri dish.

Although Koch had already developed techniques for growing bacteria on solid media prior to the invention of the Petri dish, the Petri dish represented a significant improvement in terms of practicality and efficiency. Koch recognised the importance of this innovation and supported the widespread dissemination and adoption of the Petri dish in the scientific community. The relationship between Petri and Koch was professional and collaborative, with Petri serving as a lab assistant and contributing creative ideas that led to the development of the Petri dish. Although there is no evidence of significant conflict between them, the story suggests that Koch sometimes failed to give proper credit to his collaborators, which may have influenced the perception of Petri's work in relation to the Petri dish.

The Petri dish not only simplified the cultivation process, but also opened new doors in microbiological research. With this tool, scientists were able to study the morphology, growth and behaviour of a wide range of microorganisms in a more detailed and precise way.

The influence of the Petri dish extended far beyond Koch's laboratory. Over time, it became an indispensable tool in microbiology laboratories around the world, used for a variety of purposes, from diagnosing infectious diseases to researching new drugs and treatments. Today, the Petri dish remains an essential part of any microbiologist's arsenal of tools, tangible proof of the lasting impact of the Petri and Koch collaboration on the world of science.

The collaboration between Petri and Koch was crucial to the advancement of microbiology. Together, they explored new culture techniques that would forever change the way microorganisms are studied. Picture Koch's laboratory at the Kaiserliches Gesundheitsamt in 19th century Germany, "Imperial Health Office". This institution was created to address issues related to public health in the empire. Its main purpose was to oversee and promote public health measures, including disease prevention, hygiene and medical research. A vibrant environment of discovery where brilliant minds immersed themselves in the search for innovative solutions to the scientific challenges of the time.

Julius Richard Petri (1852-1921) was a German military doctor and bacteriologist whose contributions revolutionised the field of microbiology. Born on 31 May 1852 in Barmen, Germany, Petri received his medical training at the Haiser Wilhelm Akademie für Medizin between 1871 and 1875, where he developed a particular interest in medical and microbiological sciences. Subsequently, he served as an assistant physician at the Berlin Charité and was assigned to the Kaiserliches Gesundheitsamt, where he worked alongside Robert Koch between 1877 and 1879.

During his time at the Kaiserliches Gesundheitsamt, Petri was involved in the renewal of developing microbiological techniques. He introduced innovations such as the design of new vessels and containers for collecting samples, as well as the use of sand filters. However, his most notable contribution was the creation of the Petri dish, an essential tool in the cultivation of microorganisms. The Petri dish not only simplified the cultivation process, but also opened new doors in microbiological research. With this tool, scientists were able to study the morphology, growth and behaviour of a wide range of microorganisms in a more detailed and precise way. Think of the Petri dish as a miniature garden for bacteria and fungi, where scientists can grow and observe these tiny living things. Petri designed this tool as a means to provide microbiologists with a controlled environment where microorganisms could grow and multiply in a solid culture medium. This innovation not only simplified the cultivation process, but also allowed microorganisms to be studied in more detail.

In addition to his work in microbiology, Petri also served as curator of the Hygiene Museum in Berlin and held various positions in the field of public health and hygiene. He retired as Geheimer Regierungsrat in 1900 and died in Zeitz, Germany, on 20 December 1921.

Throughout his career, Petri published numerous papers on microbiology, hygiene and laboratory techniques, including his famous article "Eine kleine Modifikation der Koch'schen Plattenverfahrens" (A small modification of Koch's plate method), where he introduced his invention to the scientific world. The influence of the Petri dish extended far beyond Koch's laboratory. Over time, it became an indispensable tool in microbiology laboratories around the world, used for a variety of purposes, from diagnosing infectious diseases to researching new drugs and treatments. Today, the Petri dish remains an essential part of any microbiologist's arsenal of tools, tangible proof of the lasting impact of the Petri and Koch collaboration on the world of science.

1.4 Development and evolution of the Petri dish culture technique Petri dishes

The Petri dish has provided a space for the growth and development of micro-organisms in the early days of micro-organism research, but with it, various techniques have also been developed, including, in principle, culture media.

These in turn are fundamental to microbiology, as they provide microorganisms with the nutrients necessary for their growth and development. To understand their importance, let's imagine a garden: just as plants need fertile soil, water and nutrients to grow, micro-organisms require a suitable medium that provides them with the nutrients they need to grow. provides carbon, oxygen, nitrogen, phosphorus, sulphur and other essential elements.

In fact, we classify them in two large groups the Macronutrients which are in general, the acronym "CHONPSK", which is used as a very useful mnemonic rule to remember the most common chemical elements present in culture media. Each letter represents one of the essential elements:

C: Carbon H: Hydrogen O: Oxygen N: Nitrogen P: Phosphorous S: Sulphur K: Potassium

Microorganisms are living organisms that need a specific combination of nutrients to grow and reproduce. These nutrients provide energy and chemical elements that are essential for the synthesis of cell molecules. When analysing the composition of cells, we find that the main components are carbon, oxygen, nitrogen, phosphorus and sulphur. To better understand the importance of these elements, let us consider their proportions in cells. Carbon, for example, makes up about 50% of the dry weight of cells. It is the main component of organic molecules, including carbohydrates, lipids, proteins and nucleic acids. Oxygen accounts for about 32% of the dry weight and is essential for cellular respiration and other metabolic reactions.

Nitrogen constitutes approximately 14% of the dry weight and is a critical component of amino acids, nucleic acids and other cellular molecules. Phosphorus, present in approximately 3%, is necessary for the synthesis of ATP, nucleic acids and phospholipids. Sulphur, although found in much smaller proportions (about 1%), is essential for the structure of some amino acids and vitamins.

When designing culture media for micro-organisms, we must provide these elements in assimilable forms. For example, heterotrophs require carbon in the form of organic compounds, such as carbohydrates and proteins, while autotrophs can use carbon dioxide. Nitrogen can be supplied in the form of ammonium (NH_4), nitrate (NO_3^-) or nitrite (NO_2^-), or via amino acids. Phosphorus is added in the form of phosphate (PO_4^{3-}) and sulphur can come from sulphur amino acids or sulphate (SO_4^{2-}). In certain cases, it is necessary to add additional amino acids or vitamins to the culture medium to meet the specific needs of certain micro-organisms that cannot synthesise these compounds on their own. This ensures that the micro-organisms have access to all nutrients necessary for optimal growth.

These elements are essential for the growth and reproduction of micro-organisms. For example, carbon and hydrogen are building blocks of organic molecules, such as carbohydrates, lipids and proteins. Oxygen is essential for cellular respiration and energy generation. Nitrogen is found in amino acids, nucleic acids and other molecules important for cell structure and function. Phosphorus and sulphur are necessary for the synthesis of molecules such as ATP and sulphur amino acids. Microelements, also known as trace elements, are chemical elements that are essential for the growth of microorganisms, although they are only required in very small quantities. These elements are vital for various cellular functions and metabolic processes.Examples of important micro-elements in culture media include:

Iron (Fe): Essential for protein synthesis and cellular respiration in many microorganisms. Lack of iron can limit bacterial growth.
Magnesium (Mg): Important for cell membrane stability and the function of numerous enzymes.
Zinc (Zn): Necessary for enzymatic activity and protein synthesis.
Manganese (Mn): involved in enzymatic reactions and necessary for the synthesis of nucleic acids.
Copper (Cu): Important for pigment synthesis and respiratory enzymes.
Molybdenum (Mo): Required for nitrogen fixation in certain bacteria.
Cobalt (Co): Essential for the synthesis of vitamin B12 in bacteria.
Nickel (Ni): Present in certain enzymes important for nitrogen fixation.

Example from elements required for make to grow micro-organisms in culture media:

Micro-organism	Culture Medium	Microelement Present in the Culture Medium
Escherichia coli	EMB (Eosin Methylene Blue)	It contains lactose and bile salts which inhibit the growth of gram-positive bacteria, allowing the growth of E. coli.
Clostridium perfringens	TSC (Sulphite-Citrate-Thioglycolate)	Contains sulphate and citrate, acting as selective agents that inhibit the growth of other microorganisms. Allows the growth of C. perfringens.
Mycobacterium tuberculosis	Löwenstein-Jensen	Contains malachite green and nutrient salts that promote the selective growth of M. tuberculosis while inhibiting the growth of other M. tuberculosis. micro-organisms.

Micro-organism	Selective Medium	Principle
Staphylococcus aureus	Mannitol Salt Agar (MSA)	High salt concentration, inhibiting the growth of most microorganisms, except for S. aureus.
Salmonella	XLD (Xylose-Lysine-Deoxycholate)	Contains xylose, lysine and deoxycholate, helping to differentiate between different Salmonella species and suppressing the growth of other bacteria. coliforms.
Candida albicans	Sabouraud Dextrose Agar (SDA)	Contains dextrose and peptone, providing the nutrients necessary for the growth of C. albicans while inhibiting the growth of bacteria.
Pseudomonas aeruginosa	Half-cytrimide	Contains cetrimide, a selective agent that inhibits the growth of other gram-positive bacteria, allowing the growth of P. aeruginosa.
Bacillus cereus	MYP (Mannitol Yolk Polymyxin)	It contains polymyxin, inhibiting the growth of gram-negative bacteria and allowing selective growth of B. cereus.
Vibrio cholerae	TCBS (Thiosulphate-Citrate-Bile-Sucrose)	It contains sucrose and bile salts, allowing the selective growth of V. cholerae while inhibiting the growth of other micro-organisms.
Listeria monocytogenes	PALCAM (Polymyxin-Acriflavin-Lithium-Ceftazidime-Aesculin-Mannitol)	Contains a combination of antibiotics and selective, allowing selective growth of L. monocytogenes while inhibiting the growth of other micro-organisms.

These microelements are required in very small amounts compared to macronutrients such as carbon, nitrogen and phosphorus, but are equally critical for the growth and survival of micro-organisms. In culture media, these elements are supplied in the form of inorganic salts or as components of other chemical compounds necessary for cell growth. Without these micro-elements, micro-organisms can experience deficiencies that affect their ability to proliferate and perform essential metabolic functions.

Initially, the media used for micro-organism cultures were liquid, then potato slices, then gelatine and finally agar. Nowadays, culture media are solidified with agar-agar, a gelatinous substance derived from seaweed, which provides a surface for the growth of bacterial colonies.

List of companies producing Microbiological Agar of interest for Petri dish culture media:

Company	Country	Product	Code of the Product
Becton, Dickinson and Company	United States	Difco™ Agar	214530
Thermo Fisher Scientific	United States	Oxoid™ Agar	BR0016B
Merck	Germany	Merck Agar	1.01688.0 500
Bio-Rad Laboratories	United States	Bio-Rad Agar	1660102
Neogen Corporation	United States	Neogen Agar	7137
VWR International	United States	VWR Agar	90001-810
Quelab	Canada	Quelab Agar	QAA100
Carl Roth GmbH + Co. KG	Germany	Carl Roth Agar	6774.1
Hardy Diagnostics	United States	Hardy Diagnostics Agar	G40
Sigma-Aldrich, now part of of Merck	United States	Sigma-Aldrich Agar	5039
Shanghai Yuanye Bio-Technology Co., Ltd.	China	Yuanye Agar	-

Qingdao Hope Bio-Technology Co., Ltd.	China	Hope Agar	-
Biokar Diagnostics	France	Biokar Agar	-
Tianjin New Sun Biochem Co., Ltd.	China	New Sun Agar	-
BioMaxima S.A.	Poland	BioMaxima Agar	-
Guangdong Huankai Microbial Sci. & Tech. Co., Ltd.	China	Huankai Agar	-
AES CHEMUNEX	France	CHEMUNEX Agar	-
Biomérieux	France	Biomérieux Agar	-
Hangzhou Uniwise International Co., Ltd.	China	Uniwise Agar	-
Guangdong Huankai Microbial Sci. & Tech. Co., Ltd.	China	Huankai Agar	-

Imagine agar as the scaffolding of a building, providing a solid structure on which microorganisms can build their communities. Like gelatin solidifying a dessert, agar in culture media provides a firm foundation for bacterial colonies to grow and develop. This process of solidifying the culture medium with agar was developed by Fanny Hesse, wife of Walther Hesse, in collaboration with the renowned microbiologist Robert Koch at the end of the 19th century. Hesse suggested the use of agar, a gel-like substance derived from seaweed, as an alternative to the gelatin gel previously used in culture media. This innovation was revolutionary as it allowed better growth and observation of micro-organisms in laboratories. Agar acts as a kind of molecular scaffold, providing a three-dimensional structure on which micro-organisms can attach and proliferate. This substance is ideal for the cultivation of bacteria and fungi because of its ability to solidify at room temperature and its resistance to degradation by microbial enzymes.

The strength of the agar in the culture medium allows scientists to clearly observe bacterial colonies and perform various tests and analyses. microbiological. In addition, its inert nature and its ability to retain moisture make it the ideal material for the cultivation of a wide variety of micro-organisms under laboratory conditions.

Thus, agar in culture media functions as the foundation upon which scientific understanding of microorganisms is built. Its ability to solidify provides a stable platform for bacterial growth, allowing researchers to explore and discover the fascinating world of microbiology.

When some micro-organisms have specific needs, such as the addition of certain amino acids or vitamins, scientists carefully calculate the amounts to maintain the overall balance of the culture medium.

Amino acids are fundamental components of proteins, which in turn play a crucial role in the structure and function of microbial cells. The inclusion of amino acids in culture media is vital to provide microorganisms with the raw materials needed to synthesise proteins and carry out a variety of metabolic processes essential for their growth and survival. Here are some reasons why amino acids are important for some microorganisms in culture media:

Protein synthesis: Amino acids are the basic building blocks of proteins. Each micro-organism needs a variety of proteins for vital functions such as metabolism, DNA replication, RNA synthesis, nutrient transport and defence against environmental stress. Without adequate amino acids in the culture medium, microorganisms cannot synthesise these essential proteins, limiting their ability to grow and survive.

Cell growth: Amino acids are necessary for cell growth and proliferation. They act as substrates for the synthesis of new proteins and nucleic acids, which are crucial for cell division and expansion of the microbial population.Adaptation to the environment: Some amino acids have specific roles in the adaptation of microorganisms to their environment. For example, cysteine and methionine are sources of sulphur that may be important for resistance to oxidative stress and the formation of disulphide bonds in structural proteins. Other amino acids, such as proline, can act as osmolytes and protect cells against desiccation and other forms of osmotic stress.

Microbial interactions: In some cases, amino acids can affect interactions between micro-organisms in a given environment. For example, the availability of certain amino acids may influence competition between different species for resources in the environment.This may have important consequences for the structure and dynamics of microbial communities.

Table of some specific amino acids needed for the growth of micro-organisms:

Micro-organism	Amino acid required	Role in the culture medium
Clostridium spp.	Cysteine	Source of sulphur for protein synthesis and coenzymes
Bacillus spp.	Proline	Precursor of protein synthesis and agent osmoprotector
Lactobacillus spp.	Proline	Precursor of protein synthesis and agent osmoprotector
Streptococcus spp.	Arginine	Energy production and precursor for energy synthesis polyamines
Lactobacillus spp.	Glycine	Precursor of protein synthesis and agent osmoprotector
Escherichia coli	Tyrosine	Source of carbon and nitrogen, and a precursor of the protein synthesis
Pseudomonas aeruginosa	Tyrosine	Source of carbon and nitrogen, and a precursor of the protein synthesis

Salmonella spp.	Methionine	Source of sulphur for the synthesis of cysteine and others biomolecules
Proteus spp.	Methionine	Source of sulphur for the synthesis of cysteine and others biomolecules
Haemophilus influenzae	Histidine	Necessary for the growth of strains that are unable to synthesise it
Bordetella pertussis	Histidine	Necessary for the growth of strains that are unable to synthesise it
Clostridium spp.	Tryptophan	Precursor of protein and metabolite synthesis secondary
Staphylococcus spp.	Tryptophan	Precursor of protein and metabolite synthesis secondary

Demanding micro-organisms are those that have specific needs for growth and survival in culture media. Among these needs are amino acids, essential components for protein synthesis and other vital cellular processes. These microorganisms, such as Corynebacterium diphtheriae, Neisseria gonorrhoeae, Yersinia enterocolitica, Neisseria meningitidis and Listeria monocytogenes, require amino acids such as methionine, lysine, valine, arginine and tyrosine, respectively, for their proper development in the laboratory. The inclusion of these amino acids in specific culture media allows the selective growth and subsequent identification of these pathogens, which is essential in the study and diagnosis of infectious diseases.

Table of Some micro-organisms needing specific amino acids for demanding micro-organisms:

Amino acid	Name of the culture medium	Micro-organism	Explanation
Methionine	Löffler medium	Corynebacterium diphtheriae	Corynebacterium diphtheriae, the causative agent of diphtheria, requires methionine for protein synthesis. Löffler's medium, which contains methionine, is used for selective cultivation.
Lysine	Mueller-Hinton medium	Neisseria gonorrhoeae	Neisseria gonorrhoeae, the causative agent of gonorrhoea, requires lysine for its growth. Mueller-Hinton medium, which contains lysine, is used for its growth. isolation and cultivation.
Valina	Cefsulodin - Irgasan - Novobiocin (CIN) medium	Yersinia enterocolitica	Yersinia enterocolitica, a pathogenic bacterium associated with intestinal infections, requires valine for its growth. CIN medium, which contains valine, is used for its cultivation and selective isolation.
Arginine	Thayer-Martin medium	Neisseria meningitidis	Neisseria meningitidis, an important pathogen causing meningitis and septicaemia, requires arginine for its metabolism. Thayer-Martin's medium, which contains arginine, is used for its metabolism. isolation and selective cultivation.

Tyrosine	Listeria Selective Agar Medium	Listeria monocytogenes	Listeria monocytogenes, a foodborne pathogen, requires tyrosine for growth. Listeria Selective Agar medium, which contains tyrosine, is used for selective culture and detection in samples. foodstuffs.

Vitamins are essential organic compounds that play critical roles in the metabolism and cell growth of micro-organisms. Although micro-organisms have the ability to synthesise many vitamins on their own, some cannot produce certain vitamins or synthesise them in insufficient amounts to meet their nutritional needs. Therefore, these vitamins must be provided in culture media to ensure optimal growth of microorganisms. Here are some common vitamins needed in microbiological culture media:Biotin (Vitamin B7): Biotin is essential for the synthesis of fatty acids and the metabolism of carbohydrates and proteins. It is needed for the synthesis of coenzymes involved in numerous metabolic reactions. Many micro-organisms require biotin in their culture media, as they cannot synthesise it in sufficient quantities.

Pantothenic acid (Vitamin B5): Pantothenic acid is a crucial component of coenzyme A, which plays a key role in the metabolism of lipids, carbohydrates and proteins. It is required for the synthesis of fatty acids and the transfer of acyl groups in numerous biochemical reactions. Niacin (Vitamin B3): Niacin is important for energy metabolism and the synthesis of NAD and NADP, coenzymes involved in electron transfer in redox reactions. It is essential for many metabolic pathways, including glycolysis and cellular respiration.Riboflavin (Vitamin B2): Riboflavin is a component of the coenzymes FMN (flavin mononucleotide) and FAD (flavin dinucleotide), which are involved in electron transport and energy production in the respiratory chain and electron transport chain. Thiamine (Vitamin B1): Thiamine is necessary for the synthesis of coenzymes involved in carbohydrate metabolism, especially in the oxidative decarboxylation of pyruvate in the Krebs cycle.

Folic acid (Vitamin B9): Folic acid is essential for the synthesis of nucleic acids (DNA and RNA) and is crucial for cell growth and division.

Vitamin B12 (Cobalamin): Vitamin B12 is necessary for the metabolism of amino acids and nucleic acids. It is essential for cell growth and nucleic acid synthesis.

Vitamin K: Vitamin K is important for the synthesis of proteins involved in blood clotting and for the activation of certain proteins involved in the metabolism of micro-organisms.

The inclusion of these vitamins in microbiological culture media ensures that microorganisms have access to the cofactors necessary for proper growth and metabolism. This is especially important in laboratory environments where microorganisms may be limited in terms of nutrient sources.

Table of some media to which Vitamin is added to grow demanding microorganisms:

Means of Cultivation	Vitamin	Micro-organism	Reason for Need
Blood Agar	NAD	Haemophilus influenzae	Haemophilus influenzae needs NAD for its growth, as it cannot synthesise this vitamin on its own. NAD is important for many metabolic reactions in the micro-organism.
Tetrathionic Medium	Folic acid	Streptococcus pneumoniae	Streptococcus pneumoniae requires folic acid for DNA and RNA synthesis. The tetrathionic medium provides folic acid for the growth of this bacterium.
Half Tinsdale	Vitamin B12	Corynebacterium diphtheriae	Corynebacterium diphtheriae needs vitamin B12 for its metabolism. The Tinsdale medium provides this vitamin to allow the selective growth of this bacterium.
Löwenstein-Jensen medium	Biotin	Mycobacterium tuberculosis	Mycobacterium tuberculosis requires biotin for growth and metabolism. Löwenstein-Jensen's medium contains biotin to support the growth of this bacterium.
Agar Sabouraud	Thiamine	Candida albicans	Candida albicans needs thiamine for its energy metabolism. Sabouraud agar enriched with thiamine provides the right conditions for the growth of Candida albicans. fungus

In the vast world of microbiology, where life unfolds on tiny scales and the mysteries of existence are revealed through the microscope, culture media are essential tools that allow us to explore and understand the secrets of microorganisms.Imagine the Petri dish as a blank canvas where microbes, like invisible artists, paint their own landscape. But what kind of paint do they use, and how does this affect their work? We are going to look at two categories of culture media: defined and complex.Defined or Simple Media are media where all the constituent elements are known both in quantity and quality, i.e. all the components are known one by one and also the quantities measured in standard units either in grams, milligrams or any other unit of interest. In a medium of this type it is possible to identify all the components on the labels without ambiguous forms and with the formula of each component.In the world of defined culture media, precision is key. Every component, from glucose to amino acids to vitamins, is dosed with surgical precision. It's like following a recipe with scientific meticulousness. Why so much precision? Because in science, control is the cornerstone. Imagine you are setting up an experiment to study how certain bacteria grow in different conditions. If you want to ensure that the results are consistent and reproducible, you need to be sure that each bacterial cell has access to the same nutrients in each repetition of the experiment. Think of a meticulous gardener who precisely measures out the exact amount of water and fertiliser for each plant. This ensures that each has an equal chance to grow and flourish. Similarly, in the world of microbiology, defined culture media act as controlled nutrition for microorganisms, allowing scientists to study their behaviour with confidence.

Examples:

Glucose: as a source of energy for microbes, it's like giving them candy to work with.

Mineral salts: are like essential building blocks for microbial cells, providing the elements necessary for their growth and function.

Amino acids and vitamins: these are like vitamins and nutritional supplements for micro-organisms, ensuring that they have everything they need to thrive.

Complex or Undefined Media These media contain various elements of which the exact formula and quantity are unknown, as they can be of such complex components that they are undefined by nature, for example, in a medium with meat extract, is it possible to know all the components of the meat? As this complexity is enormous, then it is undefined and the media, although known to be useful for the growth of micro-organisms, pass to a level where some components are undefined.

Let's look at one way of explaining the above, now, let's shift the focus to complex cultivation media. Imagine an ancient recipe passed down from generation to generation, where the secret ingredients are jealously guarded by each cook. These media are composed of a mixture of extracts of complex materials, such as yeast extract or meat extract. The interesting thing here is that we don't know all the components at play. It is like entering a mysterious forest where unknown creatures lurk in every shadow.Why opt for this uncertainty? Well, in nature, life is complex and diverse. Some microorganisms are so demanding that they need a wide variety of nutrients to thrive, and this is where complex media come into play. It's like offering a varied buffet rather than a single dish. This ensures that micro-organisms have access to a range of nutrients, which can be crucial for their growth and development.

Examples:
Yeast or meat extract: provide a variety of nutrients, such as carbohydrates, proteins, vitamins and minerals, creating a rich and complex environment for micro-organisms.

Peptone: a derivative of hydrolysed proteins, which provides a source of amino acids and peptides, essential for cell growth.
Soy or malt extract: adds a variety of additional nutrients, further broadening the spectrum of substances available to micro-organisms.

The choice between defined and complex culture media is like deciding between a precise and known recipe versus a secret and complex recipe handed down through generations. Both have their own advantages and applications, depending on the needs of the experiment and the characteristics of the micro-organisms under study. What is clear, however, is that whether we are exploring the precision of the laboratory or the complexity of the ecosystem, culture media are the fundamental tools that allow us to unravel the mysteries of the microbial world.

Preparing culture media for Petri dishes is like embarking on a historical culinary journey, where every step is crucial to grow and observe microorganisms in the laboratory. Following a meticulous recipe, each ingredient is added in the precise amount, evoking the spirit of the ancient alchemists who searched for the perfect formula to transform simple materials into something extraordinary.

As in a gastronomic journey, there are two main types of culture media: defined and complex. The former, like ancient recipes handed down from generation to generation, have a precise and known chemical composition. Meanwhile, complex media are like a renowned chef's exotic blends of ingredients, with extracts of complex materials that yield a range of nutrients and flavours. Solid media like the foundations of an ancient fortress, solidified with agar, a gelatinous substance derived from seaweed. This agar, like cement paste in the hands of a mason, provides a firm foundation on which micro-organisms can build their colonies, ensuring their growth and development. The idea of using agar in culture media came at a historic moment, when Fanny Hesse, perhaps inspired by the alchemists of old, collaborated with the renowned microbiologist Robert Koch to develop a revolutionary innovation. Her proposal to use agar as an alternative to the gelatin gel previously employed opened new doors in the science of microbiology, allowing better study and observation of microorganisms in laboratories. Agar, then, becomes the hero of this story, acting as a molecular scaffold that supports and nurtures microorganisms on their journey of growth and discovery. Its ability to solidify at room temperature and resist degradation by microbial enzymes makes it an invaluable ally in the exploration of the invisible world that inhabits our Petri dishes.

The robustness of the agar in the culture medium allows scientists to clearly observe bacterial colonies and perform various microbiological tests and analyses. In addition, its inert nature and its ability to retain The moisture content makes it the ideal material for the cultivation of a wide variety of micro-organisms under laboratory conditions.

This is how agar in culture media functions as the foundation on which scientific understanding of microorganisms is built. Its ability to solidify provides a stable platform for bacterial growth, allowing researchers to explore and discover the fascinating world of microbiology.

Depending on their function, culture media can be classified as general, selective, differential or enrichment media. General media provide the basic nutrients for the growth of a wide variety of micro-organisms. Selective media, on the other hand, favour the growth of certain microorganisms while inhibiting the growth of others. Imagine a competition in which only a few participants can advance. Differential media allow different types of micro-organisms to be identified on the basis of specific characteristics, such as the ability to ferment certain sugars or produce certain pigments. Finally, enrichment media are designed to isolate a particular type of micro-organism from a complex mixture, providing optimal conditions for its growth while limiting the growth of others. A common example of a selective culture medium is MacConkey agar, which contains crystal violet and bile salts to inhibit the growth of gram-positive bacteria and promote the growth of gram-negative bacteria, such as Escherichia coli. This medium is also differential, as it allows the identification of lactose-fermenting bacteria by producing pink colonies.In the Petri dish, culture media may vary depending on the type of micro-organism being grown and the purpose of the experiment. Here is a description of some common types of culture media used in Petri dishes:

General culture medium: Consists of a mixture of meat extract, yeast extract, peptone and agar. It provides the nutrients necessary for the growth of a wide variety of micro-organisms and is commonly used for general purposes, such as observation of the morphology of bacterial colonies.

Selective culture medium: This type of culture medium is designed to favour the growth of certain microorganisms while inhibiting the growth of others. For example, MacConkey agar is used to isolate and distinguish gram-negative enteric bacteria, such as Escherichia coli, while inhibiting the growth of gram-positive bacteria. This is achieved by adding crystal violet and bile salts to the medium. Differential culture medium: Similar to selective agar, differential agar allows distinguishing between different types of microorganisms according to specific characteristics. For example, blood agar is used to identify bacteria that are haemolytic (destroy red blood cells) and those that are not. Haemolytic colonies produce areas of haemolysis around them, making them easily identifiable.

Enriched culture medium: This type of culture medium is enriched with additional nutrients to support the growth of demanding or fastidious micro-organisms. For example, enriched blood agar is used for the cultivation of micro-organisms that require additional nutrients, such as some Streptococcus species.

Differential and selective culture media: Some culture media combine selectivity and differentiation characteristics. For example, EMB (methylene blue eosin) agar is used for selective isolation of enteric gram-negative bacteria and differentiation of lactose-fermenting bacteria. Chromogenic culture medium: This type of culture medium contains chromogenic substrates that allow visual detection of specific microorganisms. For example, CHROMagar™ agar is used for the rapid and differential identification of bacterial species, such as Escherichia coli, Salmonella spp. and Enterococcus spp. by producing colonies of different colours. Anaerobic culture medium: Some Petri dishes are designed to grow micro-organisms that grow best in the absence of oxygen. These plates may contain a reducing agent, such as thioglycolate, which removes oxygen from the medium and creates an anaerobic environment.

Transport culture medium: Used to maintain the viability of micro-organisms during transport from the sampling site to the laboratory. This type of agar generally contains ingredients that prevent overgrowth of micro-organisms during transport.

Identification culture medium: This type of culture medium is used to identify unknown micro-organisms by biochemical or serological tests. It may contain specific substrates and reagents to detect enzymatic activities or metabolic reactions characteristic of certain micro-organisms.

Antifungal culture medium: Used for the selective culture of bacteria by inhibiting the growth of fungi and yeasts. This agar may contain antifungal agents, such as chloramphenicol or cycloheximide, which suppress the growth of undesirable microorganisms.

These are just a few examples of the types of culture media used in Petri dishes. Each type of media has its own specific composition and purpose, which allows microbiologists to select the most suitable medium for your experiments according to the micro-organism you wish to culture and the objectives of your research.

Culture media are like the fertile soil in which plants grow, providing micro-organisms with the nutrients necessary for their development. Their diversity and adaptability make them essential tools in microbiological research, allowing us to study and better understand the invisible world around us. I often explain to my students that when we grow a Petri dish we must be aware of the incubation times as the microorganisms are like guests at a banquet where there is a feast of food where the microorganisms are fed in the case of bacteria if 24h pass to see their growth this is the right time for the microorganisms to grow and see their cultural characteristics in the Petri dish and it is at that time when it is possible to identify them accurately, However, a carelessness to let another day or more pass will challenge the possibility that these microorganisms finish with the feast and feed to such a degree that they grow in proportions of 2^n growing and to such a degree that they fill the plate and it is not possible to identify them accurately or at least recognise their cultural characteristics because of excess growth or because their quantity is such that they are countless, of course a trained eye can possibly recognise many of these characteristics but for purposes of didactic training, teaching, and scientific verification it is very difficult to identify them accurately or at least recognise their cultural characteristics. It is important to take the necessary precautions to properly observe the micro-organisms. And while bacteria grow in about 24 hours, fungi can be visualised in a period of 3 to 5 days where it is possible to identify their cultural characteristics. All this leads us to recognise how these micro-organisms take up a large amount of nutrients to such an extent that the population increases in relation to the amount of nutrients present.

1.5 Impact on the understanding of ancient microbiology

The invention and use of the Petri dish has had a revolutionary impact on the understanding and development of microbiology. To fully appreciate its significance, imagine a world before its existence. Scientists faced the challenge of studying micro-organisms, creatures invisible to the naked eye that could nevertheless cause devastating diseases and transform ecosystems. Without a proper tool to isolate and observe these micro-organisms, their studies were like trying to understand the nature of fish by observing a muddy lake. The Petri dish, invented by Julius Richard Petri in 1887, radically changed this situation. Petri, this German bacteriologist who worked as an assistant to Robert Koch, one of the fathers of the microbiology, he introduced this simple but ingenious vessel: a cylindrical glass dish with a tight-fitting lid. This tool allowed scientists to grow microorganisms in a solid medium, which transformed their ability to observe and study these organisms with precision.

The Petri dish allowed the isolation of individual colonies of micro-organisms, essential for the identification and study of specific bacteria and fungi. Prior to this, culture methods were imprecise and often contaminated. Imagine trying to study a single flower in a densely populated field without being able to get close or touch it. The Petri dish allowed scientists to isolate individual colonies of microorganisms, essential for the identification and study of specific bacteria and fungi. This was like having a well-organised garden where each plant grows in its own plot, allowing their characteristics to be observed without interference.

For example, the ability to grow a single species of micro-organism in a petri dish allowed scientists to study its characteristics and behaviour without interference. This was instrumental in identifying specific pathogens responsible for diseases, such as tuberculosis and cholera, and led to better diagnosis and treatment. The Petri dish played a key role in the formulation and verification of Koch's postulates. These postulates, developed by Robert Koch, set out criteria for demonstrating that a specific micro-organism causes a specific disease. Imagine having to prove that a specific spark from a forest fire was the cause of a fire that destroyed an entire forest. The Petri dish made it possible to isolate and grow bacteria on solid media, demonstrating the causal relationship between micro-organisms and disease with the same precision as a detective could trace the cause of a fire to its origin. The use of the Petri dish allowed scientists to observe and describe the morphology of bacterial colonies in great detail. Different bacteria formed colonies with distinctive characteristics, such as specific shapes, sizes, colours and textures. This is comparable to a gardener learning to identify plants not only by their flowers, but also by their leaves, stems and growth patterns. These observations were fundamental to microbial taxonomy and to the identification of pathogenic microorganisms. For example, the bacterium Staphylococcus aureus forms golden, rounded colonies, while Escherichia coli produces grey, flat colonies. These morphological differences are crucial for microbiologists in the identification and study of bacteria.

The Petri dish facilitated the development and use of various culture media. One of the most important advances was the introduction of agar as a solidifying agent. It recreates trying to grow plants in soil that constantly collapses and does not hold its shape. Agar, derived from algae, provided a solid, stable surface that did not liquefy at low temperatures, significantly improving the handling and observation of microbial cultures. With the ability to grow bacteria in Petri dishes, scientists were able to perform controlled experiments on the pathogenesis, virulence and antibiotic resistance of different bacteria. This is similar to how a doctor can study the behaviour of a virus isolated from a blood sample to develop a specific treatment. This capability facilitated the development of vaccines and treatments for various infectious diseases, improving global public health. For example, the identification and study of Mycobacterium tuberculosis in petri dishes led to the development of effective antibiotics to treat tuberculosis, a disease that had caused millions of deaths worldwide.The Petri dish allowed the standardisation of microbiological techniques, improving the reproducibility of experiments. This is crucial for scientific validation. Imagine a chef who can replicate exactly the same recipe in different kitchens around the world. Results obtained in one laboratory could be replicated in others, establishing microbiology as a rigorous scientific discipline. For example, the ability to replicate microbiological experiments in different laboratories allowed scientists to confirm discoveries and advance scientific knowledge in a collaborative and consistent manner. The simplicity and effectiveness of the Petri dish made it an essential educational tool. It facilitated the teaching of basic microbiological techniques, such as the cultivation, isolation and observation of bacteria. This is comparable to teaching young cooks the basics of cooking before moving on to more complex dishes. This hands-on approach fostered the development of skills in students and professionals, consolidating microbiology as a fundamental field of study.

The ability to identify and study specific pathogens using the Petri dish had a direct impact on medicine and public health. It enabled the development of better diagnostic methods, more effective treatments and prevention strategies for infectious diseases. This contributed to the reduction of mortality and morbidity associated with infectious diseases, improving the quality of life globally.

1.6 Historical case studies and key discoveries

The Petri dish has been instrumental in a number of landmark discoveries that have shaped modern microbiology. Here are some of the most prominent case studies and key discoveries that were made possible through the use of this tool.

Robert Koch, known for formulating Koch's postulates, used Petri dishes to isolate and culture Mycobacterium tuberculosis, the causative agent of tuberculosis, in 1882. By using solid media in Petri dishes, Koch was able to observe the specific characteristics of the bacterial colonies and perform tests that proved that this bacterium was responsible for the disease. This discovery not only confirmed the germ theory of disease, but also led to the development of effective diagnostic and treatment strategies, saving countless lives. In 1928, Alexander Fleming made one of the most famous discoveries in microbiology with the Petri dish. While was investigating Staphylococcus aureus, Fleming noticed that one of his Petri dishes had been contaminated by mould and that around the mould was an area free of bacteria. This observation led to the discovery of penicillin, the first antibiotic, which revolutionised medicine by providing an effective tool against bacterial infections. Without the ability to observe these interactions in a solid medium, this finding could have gone unnoticed.

In the 1980s, Barry Marshall and Robin Warren used Petri dishes to culture and isolate Helicobacter pylori, a bacterium implicated in most peptic ulcers and some types of gastritis. Prior to this discovery, ulcers were thought to be caused mainly by stress and diet. The use of Petri dishes allowed Marshall and Warren to isolate the bacteria and demonstrate their role in the disease, which led to the development of specific antibiotic treatments. This discovery earned them the Nobel Prize in Medicine in 2005. The Petri dish has been instrumental in studies of antibiotic resistance. For example, in the 1940s and 1950s, Joshua Lederberg used Petri dishes to study genetic recombination in bacteria and the transfer of antibiotic resistance. His work demonstrated how bacteria could exchange resistance genes, providing a fundamental understanding of the mechanisms behind antibiotic resistance. These studies have been crucial in developing strategies to combat antibiotic resistance, an increasingly serious public health problem. The Petri dish has also been instrumental in the development of cell culture techniques that have had a profound impact on cell biology and medicine. For example, HeLa cells, one of the most widely used cell lines in scientific research, were first cultured in Petri dishes in the 1950s. These cells have enabled significant advances in the study of cancer, virology, genetics and pharmacology. Cell culture techniques developed in Petri dishes have facilitated basic and applied research in a wide range of biomedical disciplines.

Fundamental studies on cell division in bacteria were also made possible through the use of Petri dishes. The discovery and understanding of binary fission, the process by which bacteria divide to reproduce, was achieved by observing bacterial colonies. in Petri dishes. This understanding has been crucial for studies of bacterial growth, population dynamics and pathogenesis.The Petri dish has been an indispensable tool in microbiology, enabling discoveries that have transformed our understanding of bacteria and their role in health and disease. From the identification of specific pathogens to the development of antibiotics and studies on bacterial resistance, advances made with the help of the Petri dish have had a profound and lasting impact on science and medicine.

CHAPTER 2
THE MODERN

Microbiological Vision: The Petri Dish in the Contemporary Era In the showcases of many laboratories around the world, we can find a large number of round plates stacked on large shelves, often unused. Some remain unused due to an abundance of supplies, while in certain colleges and universities, these essential tools are not put to proper use. I clearly remember that when I arrived at the university, this situation was repeated in several laboratories, except for the microbiology lab. In that space, the Petri dish occupied a central place, like a precious jewel, because all studies and experiments with micro-organisms were carried out from it.

The Petri dish is, without doubt, a fundamental tool in microbiology. It allows the visualisation and cultivation of microorganisms that would otherwise be invisible to the human eye. Thanks to this simple but powerful device, it is possible to observe colonies of microorganisms with their unique characteristics of shape, colour and size, facilitating the identification of various species. As I progressed in my studies, I had the opportunity to learn from several professors, each with their own approach on how to use this glass tool to advance their research. These experiences enriched my understanding and appreciation for the Petri dish. Despite technological advances and the widespread use of DNA in contemporary modern microbiology, I discovered that so-called "classical microbiology", based on the use of Petri dishes, remains a valuable and effective methodology.During my time in the lab of Dr. Francisco Yegres, a mentor and friend, I was able to delve deeper into the classical use of the Petri dish. We tested a wide range of microorganisms, from sporophytic airborne fungi to plant pathogenic fungi. We use microorganisms in the production of food and beverages such as cheese, wine, yoghurt, must, beer, malt and acetic acid. In addition, we explore industrial applications of microbiology, such as the degradation of toxic compounds in soil and water, including polycyclic aromatic hydrocarbons, and the production of bioplastics, alcohols and biodiesel, as well as the microbiological assessment of drinking water. The Petri dish became a versatile tool for detecting various bacterial and fungal diseases. Its use in such varied studies expanded my knowledge and made me realise that almost any research in microbiology can start with this simple but essential tool. The Petri dish not only facilitates observation and cultivation of micro-organisms, but also allows a deeper understanding of biological and ecological processes.

To illustrate the importance of the Petri dish in everyday life, we can compare it to a magical window that allows us to observe a whole new universe. Imagine that we are explorers of a vast unknown cosmos, where each colony of micro-organisms is a planet with its own culture and unique characteristics. Without the Petri dish, it would be like trying to explore this universe with a foggy telescope, unable to distinguish the details that differentiate one planet from another. With the Petri dish, that window becomes clearer, revealing a cosmos full of shapes and colours that tell fascinating stories about microbial life.

From an industrial microbiology perspective, the Petri dish has allowed us to optimise production processes, such as fermentation in food and beverage production. It also plays a crucial role in bioremediation, where microorganisms are used to clean up environmental contaminants. In this context, the Petri dish is like a gardening tool that allows us to select and grow the right "plants" to restore a damaged "garden", which in this case would be the environment.

For students and professionals in microbiology, the Petri dish is not just an instrument, but a symbol of discovery and understanding. As they place samples on these plates and observe the growth of microorganisms, they are participating in a scientific tradition that has provided crucial insights for more than a century. It is a reminder that while technology evolves, the fundamental tools that connect us to the basic principles of science remain invaluable.Today we cannot imagine modern contemporary microbiology without the Petri dish. Its ability to facilitate the observation, study and manipulation of microorganisms makes it an irreplaceable part of any laboratory. From basic education to advanced research, the Petri dish continues to be an essential tool that opens a window into the microbial world, enabling advances that improve our health, our industry and our environment.

Fundamentals of Contemporary Microbiology

The Petri dish is an essential and fundamental tool in contemporary microbiology, and its use remains relevant even in the age of advanced biotechnology and genomics. This simple device has revolutionised the way we study and understand the microbial world, enabling significant advances in science, medicine and industry.

The Petri Dish: A Window into Microcosm

To understand the importance of the Petri dish, we can compare it to a magical window that allows us to observe a whole new universe. Imagine we are explorers of a vast unknown cosmos, where each colony of micro-organisms is a planet with its own unique culture and characteristics. Without the Petri dish, it would be like trying to explore this universe with a foggy telescope, unable to distinguish the details that differentiate one planet from another. With the Petri dish, that window becomes clearer, revealing a cosmos full of shapes and colours that tell fascinating stories about microbial life.

The Role of the Petri Dish in Microbiology

Since its invention by Julius Richard Petri in 1887, the Petri dish has been crucial to the development of microbiology. This simple round glass or plastic dish, together with a suitable culture medium, allows the controlled growth of bacteria, fungi and other microorganisms. This facilitates direct observation and detailed study of their morphological and physiological characteristics.

Culture and Observation of Micro-organisms:

The Petri dish allows scientists to grow micro-organisms in controlled conditions, observing how they develop, reproduce and react to different stimuli. For example, when a sample of soil or water is placed in a Petri dish with a medium of In a suitable culture medium, we can observe the growth of bacterial or fungal colonies. Each colony is visible to the naked eye and can be analysed in terms of shape, colour and size.

Identification and Diagnosis:

The ability to observe the unique characteristics of microbial colonies facilitates the identification of specific species. In medicine, this is crucial for diagnosing infections. For example, in a hospital, a physician can take a sample from a patient with an infection, culture it in a Petri dish and observe the growth of colonies. The morphological characteristics and biochemical tests performed on these colonies allow the pathogen responsible to be identified and the appropriate treatment to be chosen.

Research and Development:

The Petri dish is indispensable in microbiological research. It allows scientists to conduct experiments to understand the behaviour of microorganisms under different conditions. For example, researchers can study how bacteria respond to antibiotics by placing discs impregnated with different antibiotics in a Petri dish and observing the zones of bacterial growth inhibition around the discs.

Practical Applications in Everyday Life

The importance of the Petri dish extends far beyond the research laboratory. Its practical applications affect various aspects of our daily lives.

Food and Beverage Production:

In the food industry, microorganisms grown in Petri dishes are used for the production of food and beverages. For example, fungi and bacteria used in the fermentation of products such as cheese, wine, yoghurt and beer are selected and improved through petri dish studies. This ensures products of high quality and consistency.

Industrial Biotechnology:

In biotechnology, the Petri dish is a key tool for the development of sustainable industrial processes. For example, micro-organisms capable of degrading environmental pollutants, such as hydrocarbons in contaminated soils and water, are identified and optimised using Petri dish culture techniques. This approach is crucial for bioremediation, helping to clean up and restore the environment.

Medical Research:

In medical research, the Petri dish has been instrumental in the development of new drugs and treatments. The identification and study of pathogens, such as antibiotic-resistant bacteria, are performed in Petri dishes. In addition, antibiotic sensitivity tests, which help determine the efficacy of different drugs against specific bacteria, rely on this tool.

The Petri dish and Education

For microbiology students and professionals, the Petri dish is not just an instrument, but a symbol of discovery and understanding. As they place samples in these plates and observe the growth of microorganisms, they are participating in a scientific tradition that has provided crucial insights for more than a century. It is a reminder that while technology evolves, the fundamental tools that connect us to the basic principles of science remain invaluable.In classrooms and educational laboratories, the Petri dish is an indispensable tool for teaching microbiology. It allows students to experience first-hand how microorganisms grow and respond to different conditions. This hands-on learning is essential for understanding theoretical concepts and developing technical skills. The Petri dish, invented by German bacteriologist Julius Richard Petri in 1887, is an essential tool in microbiology and has revolutionised multiple fields, including medicine and public health. Imagine a miniature garden where the seeds of bacteria and fungi can grow and reveal their secrets. This simple circular dish, usually made of glass or plastic, has become a window into the microscopic world, allowing us to observe, identify and understand the microorganisms that affect our health.

Culture of Microorganisms and Disease Diagnosis

One of the most direct and critical applications of the Petri dish in medicine is the culture of microorganisms for the diagnosis of infectious diseases. When a patient presents with symptoms of an infection, such as fever, cough or infected wounds, a sample of blood, sputum, urine or tissue can be cultured in a Petri dish. This process is similar to planting seeds in a garden and waiting for them to germinate. If pathogenic bacteria are present, they will begin to grow and form visible colonies in the dish. For example, in the case of a urinary tract infection, a urine sample can be placed in a Petri dish with a specific culture medium. After incubation, the growth of bacteria such as Escherichia coli, a common pathogen in such infections, can be observed. This method not only confirms the presence of the infection, but also It also allows antibiotic sensitivity testing, helping doctors to select the most effective treatment.

Research on New Medicines

The Petri dish is also central to the research and development of new medicines. Scientists use petri dishes to evaluate the efficacy of new antimicrobial compounds. Imagine a group of warriors (the potential antibiotics) facing off against an army of bacteria on the battlefield of the Petri dish. Those that succeed in killing or inhibiting bacterial growth are selected for further study.A notable example is the discovery of penicillin by Alexander Fleming in 1928. Fleming observed that a fungus of the genus Penicillium had contaminated one of his petri dishes and that no bacteria grew around the fungus. This simple but revolutionary observation led to the development of the first antibiotic and transformed the treatment of bacterial infections.

Food and Water Quality Control

Beyond the hospital setting, the Petri dish plays a vital role in public health through food and water quality control. Health inspectors use petri dishes to detect the presence of pathogens in food and drinking water samples. This process is essential to prevent outbreaks of food-borne diseases such as salmonellosis and listeriosis. For example, in a food processing plant, samples can be taken from surfaces and finished products and cultured in Petri dishes. If the presence of Salmonella is detected, production can be stopped and products recalled to avoid a public health crisis. This method of surveillance is crucial to maintain food safety and protect the public.

Human Microbiome Research

The study of the human microbiome - the community of micro-organisms that inhabit our bodies - has advanced significantly through the use of the Petri dish. This research has revealed that microorganisms play a vital role in our health, affecting everything from digestion to the immune system and mental health.For example, cultures of faecal samples in petri dishes have identified beneficial bacteria that could be used as probiotics to treat digestive disorders. In addition, the petri dish has been crucial for studying how imbalances in the microbiome are linked to diseases such as obesity, diabetes and inflammatory bowel diseases.

Outbreak Prevention and Control

In public health, petri dishes are essential tools for the prevention and control of epidemic outbreaks. When an infectious disease, such as cholera or tuberculosis, emerges, public health laboratories use petri dishes to identify and characterise the pathogens responsible. This allows epidemiologists to trace the source of the outbreak and implement control measures to prevent its spread.
A contemporary example is the management of outbreaks of antibiotic-resistant infections. Laboratories can use petri dishes to isolate and study resistant strains of bacteria, helping to develop strategies to control their spread and protect vulnerable populations.

Education and Training

Finally, Petri dishes are invaluable educational tools in the training of new scientists and health professionals. In teaching laboratories, microbiology and medical students use Petri dishes to learn basic culture and diagnostic techniques. This hands-on training is essential to prepare the next generation of researchers and physicians.In conclusion, the Petri dish, although simple in design, has had a monumental impact on medicine and public health. Since the From diagnosing disease to researching new treatments and protecting food safety, this humble instrument continues to be a cornerstone in our fight against disease and promotion of health. Like a well-tended garden, the use of the Petri dish allows us to cultivate knowledge and solutions that save lives and improve the quality of life around the world.

Use in the food industry and pharmaceutical industry

Quality Control and Stability Testing

During drug production, it is essential to ensure that products are free of contaminants. Petri dishes are essential for this purpose, as they allow sterility testing and evaluation of the stability of medicinal products under various storage conditions.

In quality control, Petri Dishes are used for sterility testing, where product samples are inoculated into specific culture media contained in the plates. These are incubated under controlled conditions to allow growth of any microorganisms present. The absence of microbial growth in the plates indicates that the product is sterile, while the presence of colonies suggests contamination, which could compromise the safety of the medicinal product. In addition to sterility testing, Petri dishes are used in microbial contamination testing for non-sterile products. This involves the use of selective and differential culture media that favour the growth of specific microorganisms and allow their quantification and identification. This process ensures that the levels of micro-organisms are within acceptable limits, thus ensuring the quality and safety of the product.

Environmental monitoring is also a crucial application of Petri dishes. Petri dishes are used to monitor the microbiological quality of air and surfaces in production and storage areas. Environmental samples are grown on the plates, and the resulting colonies are counted and evaluated to ensure that hygienic conditions are maintained, minimising the risk of cross-contamination in production.

In terms of stability testing, Petri dishes are used to assess how the microbial load of a medicinal product changes over time under different storage conditions, such as variations in temperature, humidity and light. During these tests, samples of the product are inoculated into the plates at various time points. Incubation and analysis of these plates allows for the detection of any changes in the microbial load, which is essential to determine the shelf life of the product and to ensure its efficacy and safety over time.

Petri dishes are also used to detect microbial degradation of medicines. By exposing product samples to stressful conditions and culturing them in the plates, scientists can identify possible microbial degradation pathways that could affect the stability of the product. This allows the storage conditions to be adjusted or the formulation to be modified to improve the stability of the medicinal product.

Finally, in monitoring the effectiveness of preservative agents, Petri dishes play a crucial role. Products containing preservatives are inoculated with specific microorganisms and cultured in the plates at different times during stability testing. Measuring the ability of preservatives to inhibit microbial growth over time ensures that the product remains safe for use throughout its shelf life.

Antibiotic Efficacy Testing

Petri dishes are essential for antimicrobial susceptibility testing. Antibiotic discs are placed on bacterial cultures and the zone of inhibition around the discs is measured to determine the effectiveness of the antibiotic against the bacteria in question.

To carry out these tests, the Petri dish is first inoculated with a uniform suspension of the bacteria being tested. Once the plate has been inoculated and the culture medium has solidified, discs impregnated with different antibiotics are placed on the surface of the agar. The plates are incubated at a suitable temperature to allow bacterial growth. During incubation, antibiotics diffuse from the discs into the culture medium and, if effective, inhibit bacterial growth around the disc. After the incubation period, the plate is observed to identify clear areas of inhibition around the discs, known as halos. The diameter of these halos are measured and compared with pre-established standards to determine the sensitivity of the bacteria to the antibiotic.A large halo indicates a high sensitivity of the bacteria to the antibiotic, suggesting that the drug is effective in inhibiting bacterial growth. On the other hand, a small halo or no halo indicates bacterial resistance, meaning that the antibiotic is not effective against that bacterial strain. These tests are crucial for selecting the most appropriate antibiotics to treat bacterial infections, helping to avoid unnecessary use of ineffective antibiotics and contributing to the fight against antimicrobial resistance. They also provide valuable information for the formulation of clinical treatment guidelines and the management of infection control programmes in hospital settings.

Environmental Monitoring

In pharmaceutical manufacturing, environmental monitoring is critical to prevent contamination. Petri Dishes are used to sample air and surfaces in production and laboratory areas, ensuring that sterility and cleanliness standards are maintained.

To sample the air, sampling devices are used that suck in air and impact it on the surface of a Petri dish with nutrient agar. This allows any micro-organisms present in the air to be deposited on the culture medium. The plates are then incubated at a suitable temperature to allow micro-organisms to grow. After the incubation period, the plates are examined for microbial colonies, which are counted and identified to assess air quality in critical production areas. In surface sampling, Petri dishes are used with techniques such as contact plates (contact agar) and sterile swabs. In the contact plate technique, the agar surface is pressed directly against the surfaces being monitored, such as bench tops, equipment or walls. Sterile swabs are passed over the surfaces and then transferred to the culture medium on the plates. The contact plates and those containing transferred swabs are incubated to allow micro-organisms to grow.The resulting colonies are counted and evaluated to determine the microbial load on the sampled surfaces. This process is essential to verify the effectiveness of cleaning and disinfection procedures, and to identify any areas that require additional attention to maintain sterility.Petri dish environmental monitoring allows pharmaceutical manufacturers to ensure that production environments comply with regulatory standards and industry best practices. This helps prevent contamination of pharmaceutical products, ensuring their safety and efficacy for consumers. In addition, continuous monitoring provides important data that can be used to improve cleaning and disinfection procedures and to respond quickly to any signs of contamination.

Research on New Treatments

Petri dishes are central to research into new treatments and therapies. These small circular surfaces, usually made of glass or plastic, allow scientists to grow micro-organisms such as bacteria, fungi and viruses in a controlled environment. Imagine a small window into the microscopic world, where researchers can directly observe the behaviour of these tiny living things and unravel the mysteries of their mechanisms of action. This is essential for developing effective strategies to combat infectious diseases, as knowing how pathogens behave and proliferate is the first step towards defeating them.

One of the most crucial applications of Petri dishes is drug sensitivity testing. This process is similar to testing different shields in a battle to see which one best resists the enemy's attacks. By exposing microorganisms to different concentrations of drugs, scientists can determine which compound is most effective and at what minimum dose can achieve results without causing undue harm to the patient. This information is vital not only to the effectiveness of treatment, but also to prevent the development of drug resistance, a growing problem in the world of public health. But Petri dishes are not limited to microbiology; their usefulness extends to the study of cellular interactions, a crucial aspect of cell and molecular biology. Cells, like people, interact with their environment and with each other in complex and fascinating ways. Using Petri dishes, scientists can observe how cells respond to various compounds and conditions, which is fundamental to developing safe and effective therapies. For example, in the field of oncology, understanding how cancer cells react to different treatments can lead to the development of personalised therapies. These therapies are tailored to the specific characteristics of a patient's tumour, significantly improving the likelihood of success and reducing side effects.

In the field of cell and molecular biology, Petri dishes are indispensable for manipulating and observing cells in a controlled environment. This control is similar to that of a chemistry lab where every variable can be adjusted and monitored. This precision allows researchers to better understand basic cellular processes, such as cell division, signalling and programmed cell death (apoptosis). This knowledge is the basis for developing therapeutic interventions that can correct or modify these processes when they become pathological. In addition, Petri dishes have been key to the development of innovative biotechnologies. Think of tissue engineering, where scientists grow cells in a Petri dish to create tissues that can be used to replace or repair damaged tissues in the human body. Or cell cloning, which allows the production of identical cells for research or treatment. These innovations have the potential to revolutionise the treatment of a wide range of diseases and medical conditions, from organ regeneration to the creation of cellular models for testing new drugs.

Petri dishes are much more than just laboratory vessels, they are indispensable tools that open a window into the microscopic world, allowing scientists to observe and manipulate organisms and cells at a level that was previously unimaginable. This detailed and controlled access is essential to advance biomedical research, developing new treatments and therapies that can save lives and improve the health of millions of people around the world. Thanks to the versatility and precision that Petri dishes offer, science continues to unlock the secrets of life at the cellular and microbial level, paving the way for future medical and biotechnological breakthroughs.

Specific Procedures and Methods

The specific procedures and methods used with Petri Dishes are essential for the detailed study of microorganisms and cells. Among these methods, the extension seeding technique, the deep seeding technique, and the use of selective and differential media stand out. Each of these methods has its own advantages and specific applications, allowing scientists to address a wide range of research questions.

Extension Sowing Technique

Spread seeding is a fundamental technique in microbiology, like spreading butter on toast to achieve a uniform coating. In this procedure, a small amount of sample is taken and spread over the surface of the agar in the Petri dish using a tool called a seeding loop. This method is especially useful for obtaining isolated colonies, which are clusters of micro-organisms descended from a single cell. Isolated colonies are essential because they allow scientists to count and analyse individual micro-organisms, which is crucial for quantitative and qualitative studies.The precision with which this technique is performed determines the clarity and separation of colonies. Good extension seeding produces well-spaced colonies that can be easily counted and characterised. This method is widely used in clinical microbiology studies to identify pathogens in patient samples, as well as in environmental research to analyse microbial biodiversity in different ecosystems.

Deep Seeding Technique

Deep seeding is another valuable technique, comparable to mixing ingredients into a homogeneous batter before baking a cake. In this technique, the sample is mixed with molten agar at a temperature that keeps the agar liquid, but does not harm the micro-organisms. This mixture is then poured into the Petri dish and allowed to solidify. This method allows the micro-organisms to be distributed throughout the depth of the medium, not just on the surface.This technique is particularly useful for counting micro-organisms in liquid samples and for detecting anaerobic bacteria, those that cannot grow in the presence of oxygen. The micro-organisms trapped in the solidified agar form colonies at different levels, which allows a better estimation of the original concentration of micro-organisms in the sample. In addition, it provides a suitable environment for the growth of anaerobes, as the solid agar excludes oxygen, creating anaerobic microenvironments.

Use of Selective and Differential Means

Petri dishes can also contain selective and differential culture media, which are essential for studying and distinguishing between different types of microorganisms. Think of these media as smart filters that not only allow the growth of certain microorganisms, but also colour or modify them for easy identification. Selective media contain agents that inhibit the growth of some microorganisms while allowing others to grow. An example is MacConkey agar, which contains bile salts and crystal violet to inhibit gram-positive bacteria, thus favouring the growth of gram-negative bacteria. In addition, this medium is differential because it contains lactose and a pH indicator. Lactose-fermenting bacteria produce acid, which changes the colour of the indicator, making it possible to distinguish between lactose-fermenting and non-fermenting bacteria. This is crucial for identifying enteric pathogens such as Escherichia coli and Salmonella. Selective and differential media are powerful tools in diagnostic and environmental microbiology. They facilitate the rapid and accurate identification of microorganisms in complex samples, which is essential for the effective treatment of infectious diseases and for water and food quality monitoring.

Broad and deep seeding techniques, together with the use of selective and differential media, are indispensable methods that enhance the use of Petri dishes in microbiological research. These methods not only enhance our ability to study and understand microorganisms, but are also essential for developing new treatments, ensuring food safety and improving public health.The Petri dish is also a versatile and essential tool in the food and pharmaceutical industry. Its ability to facilitate culture, The observation and quantification of micro-organisms makes it indispensable to ensure the safety and quality of products. Moreover, its use in research and development contributes to continuous innovation in these fields, improving both food safety and the efficacy of medical treatments.

Use in microbial ecology and biotechnology

The Petri dish is an essential tool in scientific and medical research, allowing researchers to grow and observe microorganisms in a controlled environment. Its importance lies in its versatility and ability to provide an ideal environment for the growth of bacteria, fungi, viruses and eukaryotic cells. This facilitates a wide range of studies, from basic microbiology to clinical and biotechnological applications.

Use in Microbial Ecology

In the field of microbial ecology, the Petri dish is an invaluable tool. This field of study focuses on understanding how microorganisms interact with their environment, including other organisms and the physical environment. Here, Petri dishes allow scientists to investigate several key aspects: Isolation of Microorganisms: In nature, microorganisms exist in complex communities. Petri dishes allow researchers to isolate individual species from environmental samples, facilitating the study of their specific characteristics and behaviours.

Competition and Cooperation Studies: Microbial ecology examines how microorganisms compete for resources or collaborate in food webs. Using Petri dishes, scientists can recreate these interactions in the laboratory, observing how different species affect the growth and survival of others.

Adaptation and Evolution: Petri dishes provide a controlled environment to study how microorganisms adapt to environmental changes. This may include antibiotic resistance, utilisation of new nutrients or survival under extreme conditions.

Microbial Biogeography: By culturing samples from different habitats in Petri dishes, scientists can compare microbial diversity and better understand the geographical distribution of different microbial species.

Use in Biotechnology

Biotechnology benefits enormously from the use of Petri dishes, as these plates facilitate numerous processes essential for the development of biotechnological products and advanced research:

Genetic engineering: In genetic manipulation, Petri dishes are used to clone and select genetically modified bacteria. After introducing a new gene into a bacterial cell, scientists can use Petri dishes to grow and select cells that have correctly incorporated the desired genetic material.

Secondary Metabolite Production: Many biotechnological products, such as antibiotics, vitamins and enzymes, are secondary metabolites produced by microorganisms. Petri dishes allow researchers to isolate and optimise microbial strains that produce these compounds in significant quantities.

Biofuel development: Algae and other microorganisms grown in Petri dishes are studied for their ability to produce biofuels. Scientists can select and improve strains that have a high efficiency in producing lipids or ethanol, essential components for biofuels.

Bioremediation: Petri dishes are used to select and enhance microorganisms capable of degrading pollutants. environmental. This process involves cultivating micro-organisms in the presence of toxic substances and selecting those that show a high degradation capacity, which can then be applied in environmental clean-up processes. New Drug Development: Pharmaceutical research uses Petri dishes to study the production of bioactive compounds by microorganisms. This includes the identification of new sources of antibiotics and other drugs. They also allow the testing of drug interactions with different microorganisms, facilitating the identification of potential treatments. The Petri dish is an indispensable tool in both microbial ecology and biotechnology. In microbial ecology, it allows the isolation, study and understanding of microbial interactions and adaptations in different environments. In biotechnology, it facilitates genetic engineering, the production of important metabolites, the development of biofuels, bioremediation and the discovery of new drugs. Its ability to provide a controlled and replicable environment makes it a key player in the research and practical application of microbiology and biotechnology.

Intersection with Genetic Engineering and bioinformatics

Petri dishes are an essential tool that acts as a bridge between various scientific disciplines, including genetic engineering and bioinformatics. Their ability to provide a controlled and replicable environment makes them indispensable in these fields. In genetic engineering, they enable the manipulation and detailed study of DNA and protein expression, while in bioinformatics, they provide the experimental data needed for analysis and computational modelling. The synergy between these fields, facilitated by Petri dishes, drives the advancement of biotechnology and biomedical research, enabling the development of new therapies, biotechnological products and a deeper understanding of biological systems.

Genetic Engineering

In genetic engineering, Petri dishes play a crucial role at various stages in the process of manipulating and studying genetic material:

Gene Cloning:

Petri dishes are used to grow bacteria that have been transformed with plasmids containing genes of interest. These plates allow the selection of bacteria that have successfully incorporated the plasmid, using culture media containing antibiotics. The resulting colonies can be isolated and further analysed to verify the presence and correct expression of the introduced gene.

Recombinant Protein Expression:

Once genes have been introduced into microorganisms, Petri dishes facilitate the cultivation of these cells for the production of recombinant proteins. These proteins are essential for biological research and therapeutic applications. Petri dishes allow a critical first step to select the most productive strains before scaling them up to larger volume liquid cultures.

Mutagenesis and Mutant Selection:

Petri dishes are essential in mutagenesis studies, where mutations are introduced into DNA to study their effects. Mutants are grown in these plates, allowing the observation of altered phenotypes and the selection of those with desired characteristics for further studies or biotechnological applications.

Bioinformatics

Bioinformatics, which involves the use of computational tools to analyse biological data, also benefits from Petri dishes in several indirect ways:

Genetic Sequence Analysis:

Petri dish experiments, such as cloning and mutagenesis, generate genetic sequence data that are then analysed using bioinformatics tools. Sequence comparison and mutation identification are essential for understanding genetic changes and their effects.

Modelling of Metabolic and Regulatory Networks:

Petri dish studies provide experimental data on gene expression and metabolite production. These data are integrated into computational models that simulate metabolic and regulatory networks. Petri dishes allow these models to be validated by comparing predictions with experimental results.

Algorithm and Software Development:

Petri dishes allow the generation of large data sets that are essential for developing and improving bioinformatics algorithms and software. For example, growth and gene expression data from micro-organisms under different conditions can be used to train machine learning algorithms that predict biological behaviour.

Big Data and Omics Analytics:

Omics technologies (genomics, transcriptomics, proteomics, etc.) generate large volumes of data from experiments that often start with cultures in Petri dishes. Bioinformatics is crucial to process and analyse this data, identifying patterns and relationships that are not obvious to the naked eye.

Role in epidemiological surveillance and control of diseases

Petri dishes play a crucial role in epidemiological surveillance and disease control. These basic laboratory tools are essential for identifying, studying and monitoring pathogenic microorganisms, thus facilitating the management of disease outbreaks and the implementation of effective control measures. Their ability to provide a controlled and replicable environment for the culture and study of micro-organisms makes them indispensable in the fight against infectious diseases, helping to protect public health and prevent epidemics.

Pathogen Identification

One of the most essential uses of Petri dishes in epidemiological surveillance is the identification of pathogens. Basic procedures include:

Culture of Clinical Samples:

When an infection is suspected, clinical samples (such as blood, urine, sputum or tissue) are cultured in Petri dishes with appropriate culture media. This process allows the microorganisms present in the sample to grow and form visible colonies. These colonies can be studied to identify the pathogen responsible for the infection.

Antimicrobial susceptibility testing:

Once a pathogen has been isolated in a Petri dish, antimicrobial susceptibility testing can be performed to determine which drugs will be most effective in treating the infection. This is vital for prescribing appropriate treatments and reducing the spread of drug-resistant strains.

Outbreak Monitoring

Petri dishes are also crucial for monitoring disease outbreaks:

Early Detection:

Petri dishes allow early detection of pathogens in communities and hospitals. For example, in a hospital, samples from patients with similar symptoms can be quickly cultured to identify an outbreak of a nosocomial infection. This early detection is essential to implement control measures and prevent the spread of disease.

Continuous Surveillance:

Petri dishes are used in continuous surveillance programmes to monitor the presence of pathogens in various populations. For example, in water treatment plants, Petri dishes are used to detect the presence of coliform bacteria, indicating possible faecal contamination. This regular monitoring helps to prevent outbreaks of waterborne diseases.

Epidemiological Research

In epidemiological research, Petri dishes are used to study the dynamics of disease transmission:

Traceability of Pathogens:

During an outbreak, researchers can use Petri dishes to isolate and characterise pathogens from different patients. By comparing isolates, the sources of infection can be identified and the transmission routes. This is essential for designing effective control and prevention strategies.

Antimicrobial Resistance Studies:

Petri dishes are essential for studying antimicrobial resistance. By growing bacteria from different samples and performing sensitivity tests, researchers can map the spread of resistant strains and understand the underlying mechanisms of resistance. This information is crucial for developing antimicrobial use policies and infection control programmes.

Development of Vaccines and Treatments

The role of Petri dishes in epidemiological surveillance also extends to the development of vaccines and treatments:

Vaccine Research:

Petri dishes allow pathogens to be cultured and their behaviour to be studied, which is essential for vaccine development. Researchers can attenuate pathogens grown in Petri dishes and test their efficacy as vaccines in animal models before moving on to clinical trials.

Development of New Treatments:

By studying how pathogens grow and respond to different compounds in Petri dishes, scientists can identify new drugs and treatments. This process is essential for finding effective therapies against emerging and resistant pathogens.

Impact on science education and outreach science

Petri dishes are not only an essential tool in scientific research, but also have a significant impact on science education and outreach. These simple laboratory tools allow students of all ages to understand complex scientific concepts in a practical and accessible way. Petri dishes play a crucial role in science education and outreach by providing a practical and accessible way to teach complex scientific concepts, foster scientific skills in students, engage the community in scientific activities and inspire future scientific careers. Their versatility and ease of use make them an invaluable tool for educators, scientists and popularisers alike, contributing to the advancement of science education and public participation in science.

Practical Classroom Experience

Petri dishes offer a practical and tangible way to teach science concepts in the classroom:

Basic Microbiology:

Petri dishes are ideal for teaching basic microbiology to students of all ages. Teachers can use them to demonstrate concepts such as the growth of microorganisms, colony formation and the importance of culture conditions. The Students can perform simple experiments, such as observing bacterial growth in different culture media or investigating the effectiveness of disinfectants.

Microbial Ecology:

Petri dishes can also be used to teach microbial ecology concepts. For example, students can collect environmental samples and grow microorganisms in Petri dishes to study the diversity and distribution of microbial life in different environments. This allows them to understand the importance of microorganisms in ecosystems and how they interact with their environment.

Promoting Scientific Skills

The use of Petri dishes in the classroom encourages the development of scientific skills in students:

Observation and Analysis:

By observing the growth of microorganisms in Petri dishes, students develop observation and analysis skills. They can identify different types of colonies, analyse growth patterns and draw conclusions about the conditions that favour microbial growth.

Scientific Methods:

Petri dish experiments also teach students about the scientific method. They learn how to formulate research questions, design experiments, collect data and draw conclusions based on evidence. This hands-on approach helps students internalise the fundamental principles of science.

Science Outreach and Community Engagement

Petri dishes can be used as science outreach tools to engage the community in scientific activities:

Public Science Events:

At science fairs and other science outreach events, Petri dishes can be used for interactive demonstrations. Visitors can cultivate microorganisms, observe their growth and learn about microbiology and biotechnology in a fun and accessible way.

Community Projects:

Petri dishes can also be used in community-based citizen science projects. For example, local residents can collect water, soil or air samples and grow microorganisms to study microbial biodiversity in their environment. This encourages public participation in scientific research and promotes environmental awareness.

Boosting scientific careers

The use of Petri dishes in teaching can inspire students to pursue scientific careers:

Awakening Interest in Science:

Hands-on experiences with Petri dishes can spark students' interest in science and motivate them to explore careers in related fields, such as microbiology, biotechnology or ecology.

Professional Skills Development:

The use of Petri dishes in teaching provides students with practical skills that are relevant to scientific careers. They learn laboratory techniques, data analysis and scientific communication, preparing them for future educational and professional opportunities.

CHAPTER 3

THE FUTURE

Navigating the Microbial Universe: Future Perspectives with the Petri Dish

The Petri dish, an essential tool in microbiology, is on the cusp of a technological revolution that promises to transform scientific research and medicine. Emerging trends in its technology are driving improvements in materials and designs, enabling more precise and diverse cultures. Imagine the Petri dish as a window into unseen worlds, ever clearer and wider, revealing hidden secrets of microorganisms that were previously inaccessible.

Automation and robotics are revolutionising Petri dish handling. Think of robots as tireless little helpers in laboratories, capable of handling hundreds of plates with millimetre precision, speeding up processes that used to take days or weeks. These advances not only increase efficiency, but also minimise human error, ensuring more reliable and consistent results.

Artificial intelligence (AI) and data analytics are being integrated to interpret Petri dish results faster and more efficiently. accurate. For example, imagine software that can automatically analyse microbial colonies, identifying patterns that the human eye might miss. This is comparable to having a microscope with a mind of its own, able to detect and learn from each new piece of information, enabling revolutionary discoveries in microbial behaviour and antibiotic resistance.New theoretical and conceptual models are emerging in microbiology, offering innovative ways of understanding microbial ecosystems. Think of a microbial ecosystem as a bustling city, with bacteria, viruses and fungi interacting in complex ways. New models allow us to map these interactions in greater detail, similar to how city maps have evolved from simple drawings to detailed interactive 3D plans.Exploring the human microbiota, the vast community of microorganisms that live in our bodies, is fundamental to our health. Petri dishes allow you to study these microorganisms, revealing how they influence disease and overall health. Imagine the microbiota as a secret garden in your gut, where each microbe plays a crucial role in maintaining balance. Petri dishes allow us to grow and study these microbes, developing therapies that can restore balance in the event of disease. In regenerative medicine and cell therapy, Petri dishes are like little nurseries where stem cells and tissues can grow and develop. This opens up new possibilities for treating diseases and repairing damaged tissues. For example, heart cells can be grown to repair damaged hearts, similar to how a gardener grows plants to restore a garden devastated by a storm.

However, this progress is not without its challenges. Microbiological research raises important ethical and regulatory questions, especially in genetic manipulation and the use of sensitive data. It is crucial to navigate these challenges carefully, ensuring that science advances in a responsible and ethical manner. Think of these challenges as the gatekeepers of knowledge, who must be convinced with sound ethical arguments to allow us to move forward.

The future of the Petri dish and microbiology is bright. This humble tool will continue to be central to research, driving innovations that will transform our understanding of the microbial world and its impact on our daily lives. From the fight against infectious diseases to the exploration of microbiota and tissue engineering, the Petri dish will continue to be a beacon of scientific discovery and advancement. Just as a small crystal can scatter light in a spectrum of colours, the Petri dish will scatter the knowledge in a spectrum of applications that will benefit mankind in unimaginable ways.

Emerging trends in Petri dish technology Petri Dishes

The Petri dish, an essential tool in microbiology, is on the cusp of a technological revolution that promises to transform scientific research and medicine. Emerging trends in its technology are driving improvements in materials and designs, enabling more precise and diverse cultures. Imagine the Petri dish as a window into unseen worlds, ever clearer and wider, revealing hidden secrets of microorganisms that were previously inaccessible. Automation and robotics are revolutionising Petri dish handling. Think of robots as tireless little helpers in laboratories, capable of handling hundreds of plates with millimetre precision, speeding up processes that used to take days or weeks. These advances not only increase efficiency, but also minimise human error, ensuring more reliable and consistent results. Artificial intelligence (AI) and data analytics are integrating to interpret Petri dish results more quickly and accurately. For example, imagine software that can automatically analyse microbial colonies, identifying patterns that the human eye might miss. This is comparable to having a microscope with a mind of its own, able to detect and learn from each new piece of information, enabling revolutionary discoveries in microbial behaviour and antibiotic resistance.New theoretical and conceptual models are emerging in microbiology, offering innovative ways of understanding microbial ecosystems. Think of a microbial ecosystem as a bustling city, with bacteria, viruses and fungi interacting in complex ways. New models allow us to map these interactions in greater detail, similar to how city maps have evolved from simple drawings to detailed interactive 3D plans.Exploring the human microbiota, the vast community of microorganisms that live in our bodies, is fundamental to our health. Petri dishes allow you to study these microorganisms, revealing how they influence disease and overall health. Imagine the microbiota as a secret garden in your gut, where each microbe plays a crucial role in maintaining balance. Petri dishes allow us to grow and study these microbes, developing therapies that can restore balance in the event of illness.

In regenerative medicine and cell therapy, Petri dishes, an essential tool since their invention in the late 19th century, are undergoing significant evolution thanks to technological advances. Emerging trends in their technology are transforming their design, materials and applications, boosting their use in research and diagnostics. Here we explore these trends and their impact on the field of microbiology and biotechnology in detail.

Improving Materials

The materials used in the manufacture of Petri dishes are being optimised to improve their functionality and efficiency. Traditionally made of glass and then plastic, Petri dishes are now being manufactured with more advanced materials that offer specific advantages:

Biodegradable Plastics: Concern for the environment has prompted the development of Petri Dishes made from biodegradable plastics. These plates reduce the environmental impact associated with laboratory waste and are especially important in high turnover laboratories that generate large amounts of plastic waste. Antimicrobial Materials: New materials with antimicrobial properties are being used to reduce contamination and improve asepsis in cultures. These materials help to keep samples free of unwanted contaminants, increasing the accuracy of experiments. Modified Surfaces: Petri dish surfaces are being modified to enhance cell adhesion and microbial growth. This is particularly useful in cell culture where cell adhesion to the surface is crucial to the success of the experiment. For example, coatings with specific proteins that promote stem cell adhesion are being used for regenerative medicine research.

Innovations in Design

Petri dish design is also evolving to adapt to new needs and applications. These innovations are improving the functionality and versatility of the plates:

Multi-Well Plates: Traditional Petri Dishes are being supplemented with multi-well plates, which allow multiple experiments to be performed simultaneously in a single plate. This is especially useful in high-throughput studies where multiple wells are needed. conduct many parallel tests, such as in pharmaceutical research for drug testing.

High Density Microplates: For research requiring large numbers of small samples, high density microplates are gaining popularity. These plates allow thousands of experiments to be performed in parallel, facilitating large-scale studies such as screening libraries of chemical compounds for new drugs.

Custom Format Plates: With the help of 3D printing, it is possible to manufacture Petri dishes with custom formats adapted to specific research needs. For example, plates with microchannels and special compartments are being developed for studies of fluid dynamics and microbial behaviour in simulated environments.

Embedded Monitoring Technology

The integration of monitoring technologies within Petri dishes is revolutionising the way microbiological experiments are performed and analysed: Integrated Sensors: Sensors that can measure parameters such as pH, temperature, and oxygen concentration are being incorporated directly into the Petri dishes. This allows monitoring of the culture conditions in real time without the need to interrupt the experiment. For example, integrated pH sensors can help adjust culture medium conditions to optimise cell growth. LED Illuminated Plates: The incorporation of LED illumination in Petri dishes is allowing optical stimulation of cultures, which is particularly useful in photobiology studies and in optimising culture conditions for photosynthetic organisms. Controlled lighting can simulate day/night cycles to study circadian rhythms in microorganisms. Internet Connected Plates: With the advancement of the Internet of Things (IoT), Petri dishes are being equipped with internet connectivity capabilities, enabling remote data collection and analysis. This facilitates continuous monitoring and real-time data analysis, even from remote locations, improving efficiency and collaboration in multi-centre research.

Customisation and Scalability

The ability to customise and scale Petri dishes for various applications is opening up new frontiers in research:

Modular Petri **dishes:** Modular Petri dishes allow researchers to combine different modules in a single dish, by adapting the culture environment to your specific needs. This flexibility is key in complex experiments that require multiple experimental conditions simultaneously.

Microbioreactors: Petri dishes are evolving into microbioreactors, which allow precise control of culture conditions, such as agitation and aeration, in a compact format. These devices are revolutionising bioproduction and biotechnology research, facilitating the scalability of biological processes.

Interdisciplinary Applications

New Petri dish technologies are extending their applications to interdisciplinary fields: Tissue engineering: In tissue engineering, Petri dishes with three-dimensional scaffolds are enabling the cultivation of complex tissues that more closely resemble natural tissues. This is crucial for the development of artificial organs and tissue regeneration therapies.Microbial Ecology: Petri dishes are being used in microbial ecology studies to simulate and analyse microbial interactions in natural environments. For example, specific microhabitats can be recreated to study how different microbial species interact with each other and with their environment. Biofilm studies: The formation of biofilms, which are communities of micro-organisms that adhere to surfaces, is being studied in specially designed Petri dishes to allow detailed visualisation and analysis of these structures. This has important applications in the investigation of persistent infections and antibiotic resistance.

Emerging trends in Petri dish technology are revolutionising microbiological and biotechnological research. With improvements in materials, innovations in design, integration of monitoring technology, customisation and scalability, and interdisciplinary applications, these humble tools are poised to open new frontiers in science and medicine, enabling discoveries that will transform our understanding of the microbial universe and its impact on health and the environment. Petri dishes are like tiny nurseries where stem cells and tissues can grow and develop. This opens up new possibilities for treating diseases and repairing damaged tissues. For example, heart cells can be grown to repair damaged hearts, similar to how a gardener grows plants to restore a garden devastated by a storm.However, this progress is not without its challenges. Microbiological research raises important ethical and regulatory issues, especially in genetic manipulation and the use of sensitive data. It is crucial to navigate these challenges carefully, ensuring that science moves forward in a responsible and ethical manner. Think of these challenges as the gatekeepers of knowledge, who must be convinced with sound ethical arguments to allow us to move forward.The future of the Petri dish and microbiology is bright. This humble tool will continue to be central to research, driving innovations that will transform our understanding of the microbial world and its impact on our daily lives. From the fight against infectious diseases to the exploration of microbiota and tissue engineering, the Petri dish will continue to be a beacon of scientific discovery and advancement. Just as a small crystal can disperse light into a spectrum of colours, the Petri dish will disperse knowledge into a spectrum of applications that will benefit humanity in unimaginable ways.

Advances in automation and applied robotics

Advances in automation and robotics are revolutionising the use of Petri dishes, taking microbiological research to unprecedented levels of accuracy, efficiency and scalability. The implementation of these advances is optimising multiple aspects of microbiological research. of the experimental process, from sample handling to data analysis, facilitating faster and more accurate discoveries.

Sample Handling Automation

Automation of sample handling is one of the most significant advances in the use of Petri dishes. Automated systems can perform repetitive tasks with high accuracy and consistency, reducing the risk of human error and increasing laboratory productivity.Automatic Dispensing Systems: Automatic dispensing robots can load and dispense culture media and samples into Petri dishes with high precision. This is especially useful in high-throughput studies where the preparation of hundreds or thousands of samples is required. For example, in a pharmacological screening, a robot can prepare multiple dilutions of a compound and apply them to different plates, ensuring consistency and reducing the time needed for manual preparation.Plate Transfer and Manipulation: Specialised robots can transfer and manipulate Petri dishes between different workstations within an automated laboratory. These robots can, for example, move plates from a seeding station to an incubator and then to an analysis station, all without human intervention. This does not not only increases efficiency but also enables safe handling of hazardous or contaminating samples.

Automated Incubation and Monitoring

The automation of incubation and culture monitoring is another crucial advance. These systems can maintain optimal environmental conditions and perform continuous monitoring, ensuring proper culture growth and allowing timely interventions if necessary.

Automated Incubators: Automated incubators can control and adjust parameters such as temperature, humidity and CO_2 concentration. In addition, these incubators are equipped with continuous monitoring systems that record real-time data. For example, in the study of bacteria that require specific oxygen conditions, an automated incubator can dynamically adjust oxygen levels to optimise growth.

Real-Time Monitoring Systems: Integrating sensors into Petri dishes allows continuous monitoring of critical parameters, such as pH and gas concentration. These systems can automatically alert researchers to any deviations from the set parameters, allowing a quick and accurate response. For example, in stem cell cultures, a pH sensor can detect changes in the culture medium and adjust conditions to maintain an optimal environment for cell growth.

Automated Analysis and Processing

The automation of Petri dish data analysis and processing is transforming the way experimental results are interpreted. Automated systems can analyse large volumes of data faster and more accurately than manual methods. Automated Imaging and Analysis Systems: High-resolution cameras and automated image analysis systems can capture and analyse microbial colonies in Petri dishes. These systems use advanced algorithms to identify and count colonies, measure colony size and analyse colony morphology. For example, in antibiotic resistance studies, an automated analysis system can quickly assess the effectiveness of different antibiotics by measuring bacterial growth in the presence of these compounds.

Data Analysis Software: Data analysis software is being integrated with automated systems to process and interpret crop results. These programmes can perform complex statistical analyses and generate detailed reports, helping researchers to better understand the data obtained. For example, in a microbial genetics study, software can correlate growth patterns with genetic variations, providing valuable insights into the function of specific genes.

Collaborative Robots (Cobots)

Collaborative robots, or cobots, are designed to work alongside humans, combining the precision of automation with the flexibility of manual labour. These cobots are playing an increasingly important role in laboratories.

Assistance in Complex Experiments: Cobots can assist researchers in performing complex experiments that require a high degree of precision. For example, a cobot can help perform microinjections into single cells, a task that requires an extremely steady and precise hand.

Flexibility and Adaptability: Unlike traditional robots, which are programmed to perform specific tasks, cobots are highly adaptable and can be easily reprogrammed to perform a variety of tasks. This makes them ideal for research environments where needs can change rapidly. For example, a cobot can be used one day to dispense culture media and the next to perform image analysis. To illustrate the impact of these advances, consider a practical example in the field of infectious disease research. At a laboratory dedicated to the study of antibiotic-resistant pathogens, an automated system can handle Petri dish preparation, bacterial seeding, incubation under controlled conditions, and analysis of results, all without human intervention. Dispensing robots prepare accurate dilutions of antibiotics and apply them to the plates, while automated incubators maintain optimal conditions for bacterial growth. At the end of the experiment, an image analysis system captures and analyses the results, identifying resistant colonies and generating data for further analysis. This level of automation not only speeds up the process, but also ensures the consistency and reproducibility of the experiments, crucial elements for scientific research.

Advances in automation and applied robotics are transforming the use of Petri dishes in microbiological research. Automation of sample handling, automated incubation and monitoring, automated data analysis and processing, and the integration of collaborative robots are increasing the efficiency, precision and scalability of experiments, opening up new possibilities for scientific discovery and innovation. These technologies not only enhance the ability of researchers to perform more complex and larger-scale experiments, but also ensure that results are more accurate and reproducible, advancing knowledge in microbiology and biotechnology.

Integration of artificial intelligence and analysis of data

The integration of artificial intelligence (AI) and data analytics into microbiological research is revolutionising the use of Petri dishes. These advances are transforming the way microbiological data is collected, processed and analysed, enabling faster, more accurate and deeper discoveries. AI and data analytics not only optimise laboratory efficiency, but also expand the frontiers of scientific knowledge.

Image Analysis Automation

One of the most significant advances is the automation of image analysis of Petri dishes using AI. Microbial colonies, which were previously counted and analysed manually, can now be examined with unprecedented accuracy and speed. Pattern Recognition: AI algorithms are able to recognise and classify microbial colonies based on their morphology, colour and size. For example, AI software can analyse a Petri dish and distinguish between colonies of different bacterial species, something that can be difficult even for a human expert. This is particularly useful in microbial diversity studies, where accurate species identification is crucial. Automatic Quantification: AI can automatically count the number of colonies present on a plate, eliminating human error and improving the reproducibility of experiments. In bacterial growth studies, for example, AI can provide accurate quantitative data on growth rate and colony density, facilitating more robust comparisons and analysis.

Predictive Analytics and Modelling

The ability of AI to perform predictive analytics and modelling is transforming microbiological research. Predictive models can anticipate the behaviour of microorganisms under different conditions, allowing researchers to design more efficient and targeted experiments. Microbial Growth Models: AI can develop complex models that predict the growth of microorganisms as a function of various parameters such as temperature, pH and nutrient availability. These models are invaluable for optimising culture conditions and maximising productivity in biotechnology studies. For example, in biofuel production, AI can help identify optimal conditions for the growth of ethanol-producing microorganisms. Antibiotic Sensitivity Analysis: Using large volumes of historical data, AI can predict how different bacterial strains will respond to various antibiotics. This allows for a more targeted approach in developing new antibacterial treatments and fighting antibiotic resistance. For example, in a hospital, AI can analyse data from previous infections and predict the likelihood of resistance to certain antibiotics, thus guiding therapeutic decisions.

Big Data and Omics Analytics

The era of big data is providing huge amounts of information that AI can process and analyse to gain deep insights. This is especially relevant in omics studies, where the amount of data generated can be overwhelming.

Genomics and metagenomics: In genomic and metagenomic studies, AI can analyse DNA sequences to identify genes, discover new species and understand the structure of microbial communities. For example, in metagenomics of environmental samples, AI can assemble and annotate whole genomes from fragmented sequences, revealing the biodiversity and ecological functions of the microorganisms present. Proteomics and Metabolomics: AI is also playing a crucial role in proteomics and metabolomics, analysing large assemblies data to identify proteins and metabolites and understand their roles in biological processes. In disease studies, for example, AI can identify protein biomarkers that distinguish between healthy and diseased tissues, facilitating diagnosis and the development of personalised therapies.

Optimisation of Experimental Processes

AI is optimising experimental processes in laboratories, from planning to execution and analysis.Design of Experiments: AI can aid in the design of experiments by optimising parameters and reducing the number of trials needed to obtain meaningful results. Using techniques such as design of experiments (DOE) and Bayesian optimisation, AI can suggest the most promising experimental conditions, saving time and resources. For example, in a culture media optimisation study, AI can identify the nutrient combinations that maximise cell growth at the lowest cost.Workflow Automation: AI can coordinate and automate complex laboratory workflows, integrating different equipment and processes. An AI-powered laboratory management system can automatically schedule equipment usage, coordinate sample preparation, and ensure that data is collected and analysed. in a consistent manner. This is particularly useful in high-throughput laboratories handling large sample volumes.

To illustrate these future developments, consider a laboratory researching new antibiotic therapies. Using AI, the lab can quickly test the efficacy of different compounds against resistant bacterial strains. An automated system first seeds bacteria in Petri dishes and applies various concentrations of the compounds. The AI then analyses the images from the plates to quantify bacterial growth and determine the effectiveness of the treatments. With this data, a predictive model can suggest modifications to the chemical structures of the compounds to improve their efficacy, significantly speeding up the process of developing new antibiotics.

Integration of AI in Epidemiological Surveillance

AI is also revolutionising epidemiological surveillance and disease control. By integrating data from various sources, such as hospitals, laboratories and public databases, AI can identify disease outbreaks in real time and predict their spread.Early Outbreak Detection: AI can analyse disease case data and detect patterns that indicate the onset of an outbreak, enabling a rapid and effective response. For example, during an pandemic, AI can track the spread of the disease and help authorities implement more effective containment measures. Disease Spread Prediction: Using advanced epidemiological models, AI can predict how a disease will spread in different scenarios. This helps public health planners prepare resources and make informed decisions. In the management of an infectious disease, for example, AI can predict peaks in cases and demand for hospital beds, facilitating the efficient allocation of medical resources.

The integration of artificial intelligence and data analytics into microbiological research is transforming the use of Petri dishes. From automated image analysis to predictive modelling and omics analysis, AI is enabling significant advances in the understanding and control of microorganisms. These advances not only improve the efficiency and accuracy of research, but also open up new possibilities for scientific discovery and innovation in microbiology and biotechnology. The combination of AI and data analytics is leading microbiology into a new era of rapid and profound discovery, with potentially transformative impact on human health and well-being.

New theoretical and conceptual models on microbiology

Microbiology is evolving rapidly thanks to the introduction of new theoretical and conceptual models that are redefining our understanding of microorganisms and their interactions with the environment. In the future, the integration of these models with the use of the Petri dish will enable significant advances in biotechnology, medicine and ecology. This process of technological evolution and convergence can be visualised in several key phases:

Expansion of Microbial Ecology Models

In the near future, microbial ecology models will be applied more precisely to understand the complex interactions between different microbial species and their environment. For example, microbial food web models will be instrumental in manipulating microbiomes, improving soil health and increasing agricultural productivity. The Petri dish will allow the isolation and study of feeding relationships between different micro-organisms, facilitating experiments to validate these food webs.

Advances in Understanding Microbial Interactions

By integrating models of competition and cooperation, scientists will be able to decipher how bacteria compete for limited resources and develop symbiotic relationships. The Petri dish will continue to be a crucial tool for studying these phenomena in vitro, providing a controlled environment to observe and analyse microbial interactions in detail.

Microbial Evolution and Antibiotic Resistance

In the near future, models of adaptive evolution and horizontal gene transfer (HGT) will be essential for understanding and predicting the evolution of antibiotic resistance. The Petri dish will play a vital role in these studies, allowing scientists to directly observe how microbial populations evolve under different environmental stresses, such as antibiotic exposure.

Metabolic and Bioenergetic Optimisation

Models of metabolic networks and bioenergetics will improve our understanding of how micro-organisms obtain and use energy. In futuristic laboratories, the Petri dish will be used to experiment with specific culture conditions, optimising the production of biofuels and other compounds of biotechnological interest.

Models Applied to the Human Microbiota

For human gut microbiota, models of microbial ecology and evolution will help to understand how diet and antibiotics affect microbiota composition and function. With the Petri dish, researchers will be able to grow and study gut microbes under controlled conditions, facilitating the development of probiotics and prebiotics based on accurate, modelled data.

The complexity of microbial systems will require the integration of experimental data with theoretical models. This challenge will be met through interdisciplinary approaches combining biology, mathematics, computer science and engineering. The Petri dish will remain an indispensable tool, facilitating experimentation and validation of models that allow scientists to more accurately navigate the vast microbial universe.

Microbiology is evolving rapidly thanks to the introduction of new theoretical and conceptual models that are redefining our understanding of microorganisms and their interactions with the environment. These models not only extend basic knowledge but also have practical applications in biotechnology, medicine and ecology. With the use of the Petri dish, these models can be validated and explored experimentally, offering a powerful tool to advance multiple areas of microbiological knowledge.

Evolutionary Models: Evolutionary Networks

Horizontal Gene Transfer (HGT) models explain how genes can be transferred between different microbial species, accelerating evolution. In the Petri dish, this phenomenon can be studied by looking at the transfer of antibiotic resistance genes between bacteria. Phylogenomic models use genomic data to build phylogenetic trees and understand evolutionary relationships between microorganisms. Petri dishes allow the isolation and sequencing of different bacterial strains, facilitating the collection of data for these models. Coevolution Networks study how evolutionary interactions between different microbial species and their hosts affect evolution, and can be explored in the Petri dish by co-culturing microorganisms and host cells.

Population Dynamics Models

Competition Models and Predator-Prey Models are fundamental to understanding microbial ecological dynamics. For example, interactions between predator bacteria and their prey can be observed in real time in a Petri dish, providing valuable data to validate these models.

Theoretical-mathematical models

Population Dynamics Models, using Ordinary Differential Equations (ODEs) and Difference Equations, allow modelling the growth and interaction of microbial populations. The Petri dish is essential to perform experiments that provide the data needed to feed these models. Game theory models, such as evolutionary games and prisoner's dilemmas, applied to microbiology, can study stable evolutionary strategies such as cooperation and competition. These models are validated by observations of microbial behaviour in controlled Petri dish environments.

Complex Network Models

Interaction Network Analysis and Co-Occurrence Networks study the complex interactions between different microbial species in a community. The Petri dish allows the isolation and detailed observation of these interactions at the microscale, providing an ideal platform to validate these models.

Ecological Models

Metacommunity Models and Ecophysiology Models explore the dynamics of microbial populations distributed in different habitats and their participation in biogeochemical cycles. Petri Dishes, by simulating different environmental conditions, allow for observe how microbial communities respond to changes in their environment.

Metabolic and Bioenergetic Models

Flux Balance Analysis (FBA) Models and Bioenergetics Models analyse the metabolic behaviour and energy efficiency of microorganisms. The Petri dish is used to grow microorganisms in different conditions, allowing the collection of experimental data needed for these models.

Models of Microbial Interactions

Quorum Sensing Models explain how microorganisms use chemical signals to coordinate group activities, and can be studied by observations of microbial communication in Petri dishes. Symbiosis Models, which describe mutualistic and parasitic interactions, can also be investigated using co-cultures in Petri dishes.

Resilience Systems Models

Antibiotic Resistance Models and Biofilm Resistance Models are essential to understand how bacteria develop and maintain resistance. The Petri dish allows the study of biofilm formation and the spread of resistance genes under controlled conditions.

Models of Human-Microbe Interactions

Human Microbiota Models and Infection and Pathogenesis Models analyse the interactions between microorganisms and the human host. Petri dishes are crucial for culturing and studying microbes in conditions that simulate the human environment, providing data for these models. The future integration of these theoretical and conceptual models with the use of the Petri dish promises to revolutionise our understanding of microorganisms. These advanced models are providing new ways to understand

and predict the behaviour of microorganisms, and are driving innovations in areas such as biotechnology, medicine and ecology. As these models continue to evolve, they will continue to play a crucial role in expanding our knowledge and the practical application of microbiology. The Petri dish will remain an indispensable tool, facilitating experimentation and validation of models that allow scientists to more accurately navigate the vast microbial universe.

Exploring the human microbiota and its relationship with health

The Petri dish has been and remains a fundamental tool in microbiome research and the exploration of the human microbiota, especially when combined with "omics" technologies such as genomics, metagenomics, transcriptomics and proteomics. These disciplines allow the comprehensive analysis of micro-organisms present in biological samples, including those obtained from the human microbiota, such as faeces, saliva, skin and other tissues. Exploration of the human microbiome has revealed a fascinating network of microorganisms that coexist in our bodies, playing key roles in health and disease. The Petri dish is used in this research to culture and analyse microorganisms at the individual level, providing a detailed understanding of their physiology, metabolism and behaviour. For example, bacterial cultures obtained from faecal samples can be grown in Petri dishes and then analysed to identify and characterise the species present, their relative abundance and metabolic capabilities.In addition, the combination of the Petri dish with next-generation sequencing techniques has made it possible to study the human microbiome in a comprehensive manner. Metagenomics, for example, sequence the DNA present in a sample, making it possible to identify not only culturable bacteria in the Petri dish, but also microorganisms that cannot be cultured under laboratory conditions. This provides a more complete and accurate picture of the microbial diversity present in the human body. Exploration of the human microbiota and its relationship to health has revealed important links between microbial composition and a variety of medical conditions, ranging from gastrointestinal diseases such as irritable bowel syndrome to metabolic disorders such as obesity and diabetes. The Petri dish plays an essential role in this research by allowing the cultivation and analysis of specific microorganisms associated with different health or disease states.For example, Petri dishes can be used to isolate and culture specific bacteria suspected of being involved in a particular disease. Sequencing studies can then help to genetically characterise these bacteria and better understand their functions and effects in the human body. This may lead to the development of new therapies specifically targeting the microbiota, such as probiotics and prebiotics, which can modulate microbial composition to promote health. The Petri dish plays a central role in microbiome research and the exploration of the human microbiota by allowing the culture, isolation and detailed analysis of microorganisms. When combined with "omics" technologies, it provides invaluable information on microbial diversity and its impact on human health. This opens up new avenues for the development of therapies and treatments aimed at modulating the microbiota to improve health and prevent disease.The Petri dish has been and remains a fundamental tool in microbiome research and the exploration of the human microbiota, especially when combined with "omics" technologies such as genomics, metagenomics, transcriptomics and proteomics. These disciplines allow the comprehensive analysis of micro-organisms present in biological samples, including those obtained from the human microbiota, such as faeces, saliva, skin and other tissues. Exploration of the human microbiome has revealed a fascinating network of microorganisms that coexist in our bodies, playing key roles in health and disease. The Petri dish is used in this research to culture and analyse microorganisms at the individual level, providing a detailed understanding of their physiology, metabolism and behaviour. For example, bacterial cultures obtained from faecal samples can be grown in Petri dishes and then analysed. to identify and characterise the species present, their relative abundance and metabolic capabilities. In addition, the combination of the Petri dish with next-generation sequencing techniques has made it possible to study the human microbiome in a comprehensive manner. Metagenomics, for example, sequences the DNA present in a sample, making it possible to identify not only culturable bacteria in the Petri dish, but also microorganisms that cannot be cultured under laboratory conditions. This provides a more complete and accurate picture of the microbial diversity present in the human body.Exploration of the human microbiota and its relationship to health has revealed important links between microbial composition and a variety of medical conditions, ranging from gastrointestinal diseases such as irritable bowel syndrome to metabolic disorders such as obesity and diabetes. The Petri dish plays an essential role in this research by allowing the cultivation and analysis of specific microorganisms associated with different health or disease states. For example, Petri dishes can be used to isolate and culture specific bacteria suspected of being involved in a particular disease. Sequencing studies can then help to genetically characterise these bacteria and better understand their functions and effects in the human body. This may lead to the development of new therapies specifically targeting the microbiota, such as probiotics and prebiotics, which can modulate microbial composition to promote health. The Petri dish plays a central role in microbiome research and the exploration of the human microbiota by allowing the culture, isolation and detailed analysis of microorganisms. When combined with "omics" technologies, it provides invaluable information on microbial diversity and its impact on human health. This opens up new avenues for the development of therapies and treatments aimed at modulating the microbiota to improve health and prevent disease.

Potential in regenerative medicine and cellular therapy

The Petri dish continues to be an indispensable tool in biomedical research, especially in the context of new trends and potential in regenerative medicine and cell therapy. Its versatility and ease of use make it invaluable in the culture and manipulation of cells, as well as in the observation of their behaviour in controlled environments. In the field of regenerative medicine, the Petri dish is used to grow and expand stem cells, a crucial step in the production of tissues and organs for transplantation. Stem cells can be cultured in Petri dishes together with growth factors and other components necessary for their differentiation into specific cell types. This allows the generation of functional tissues that can be used to repair or replace damaged tissues in patients with degenerative diseases, traumatic injuries or congenital defects.

In addition, the Petri dish is central to the research and development of cell therapies, which involve the delivery of live cells to treat disease. For example, in the field of cancer immunotherapy, immune cells can be grown in Petri dishes, genetically modified to enhance their ability to recognise and destroy cancer cells, and then administered to patients. The Petri dish facilitates the expansion and manipulation of these cells, as well as the monitoring of their viability and function prior to administration.

In addition, in cell therapy, the Petri dish is used to evaluate the efficacy and safety of new cell therapies in preclinical models. For example, genetically modified cells can be cultured in Petri dishes and then transplanted into animal models to study their effectiveness in tissue regeneration or suppression of diseases. This allows cell therapies to be optimised prior to their application in human clinical trials. The Petri dish is also crucial in regenerative medicine and cell therapy research to better understand the mechanisms underlying tissue regeneration and immune response. For example, stem cells cultured in Petri dishes can be subjected to different stress conditions or stimuli to study how they respond and differentiate into different cell types. This provides valuable information about the biological processes involved in tissue regeneration and repair.

New trends and potential in regenerative medicine and cell therapy by facilitating cell culture, expansion and manipulation, as well as research into the mechanisms underlying tissue regeneration and immune response. Their continued use in combination with other innovative technologies is driving significant advances in the field of regenerative medicine and biology, with the ultimate goal of improving the health and quality of life of patients.

Ethical and regulatory challenges in research microbiology

As we advance microbiological research using the Petri dish, ethical challenges arise that must be addressed to ensure that it is conducted in a responsible and respectful manner. These ethical challenges are crucial to maintain a balance between scientific progress and respect for human rights and values. Some of the ethical challenges ahead in microbiological research using the Petri dish include:

Genetic manipulation: With the advent of technologies such as gene editing, such as CRISPR-Cas9, there are ethical concerns about the genetic modification of microorganisms for various purposes, such as the creation of genetically modified organisms (GMOs) or the alteration of pathogens to increase their virulence. It is essential to establish clear regulations and standards to guide research in this field and ensure that it is carried out in an ethical and safe manner. Bioterrorism and biosecurity: The misuse of micro-organisms for malicious purposes poses significant ethical challenges in microbiological research. The Petri dish, as a fundamental tool in the study of micro-organisms, could be used inappropriately to develop biological weapons or cause deliberate harm.It is therefore necessary to implement stringent biosafety measures and to promote awareness of the risks associated with microbiological research.

Privacy and consent: In research involving human biological samples, such as the study of human microbiota, it is crucial to respect privacy and obtain informed consent from participants. This involves ensuring the confidentiality of individuals' genetic and microbiological information, as well as obtaining their consent for the use of their samples for research purposes. The Petri dish is used in this context to grow and analyse micro-organisms present in human biological samples, highlighting the importance of addressing these ethical issues appropriately.

Equity and access: Equitable access to the benefits of microbiological research is another important ethical challenge. It is essential to ensure that scientific advances and therapies developed from Petri dish research are available and accessible to all people, regardless of their socio-economic or geographic background. This requires an ethical approach to resource allocation and the promotion of equity in access to care.

Environmental impact: Microbiological research can also have a significant impact on the environment, especially in terms of the release of genetically modified micro-organisms or the overuse of antibiotics that may contribute to bacterial resistance. It is essential to consider the potential environmental impacts of research and take measures to mitigate any negative effects on natural ecosystems.

Addressing these ethical challenges effectively will require collaboration between scientists, regulators, policy makers and society as a whole. It is essential to promote a culture of ethical responsibility in microbiological research and to ensure that it is conducted in a transparent manner, respecting fundamental ethical principles and protecting human and environmental well-being.The regulatory challenges in microbiological research using the Petri dish are diverse and complex, ranging from biosafety to the protection of human health and the environment. Some of these challenges are detailed below:

Biosafety Level: Microbiological research with the Petri dish often involves the handling of microorganisms that may pose risks to human health and the environment. It is crucial to properly classify microorganisms according to their pathogenicity and level of risk, and to apply safety protocols. according to the biosafety levels established by the regulatory authorities.

Genetic manipulation: The use of genetic engineering techniques, such as genetic modification of micro-organisms, poses regulatory challenges in terms of safety and ethics. It is necessary to establish clear regulations governing the genetic manipulation of micro-organisms and to ensure that biosafety and bioethical standards are met in research.

Culture media quality control: Culture media used in the Petri dish must meet stringent quality standards to ensure accurate and reproducible results. Regulatory challenges include standardisation of media components, detection and prevention of microbial contamination, and verification of media efficacy in growth and differentiation of microorganisms. Antimicrobial resistance: Excessive and inappropriate use of antibiotics in microbiological research can contribute to the development and spread of antimicrobial resistance. Regulations are needed to promote the responsible use of antibiotics in research and to encourage the development of alternative strategies for the control of pathogenic microorganisms. Preservation of biological samples: Proper preservation of biological samples used in microbiological research is critical to ensure their integrity and long-term utility. Regulatory challenges include the establishment of standard protocols for the collection, processing, storage and transport of biological samples, as well as the ethical and legal management of information associated with these samples. Environmental protection: Microbiological research carries the risk of accidental release of genetically modified micro-organisms or other biological agents that could have negative impacts on the environment. Regulations are needed to mitigate these risks and promote environmentally sustainable practices in microbiological research. Addressing these regulatory challenges will require close collaboration between researchers, academic institutions, government regulatory agencies and the scientific community as a whole. It is essential to develop robust and flexible regulatory frameworks that promote cutting-edge scientific research while protecting the safety and well-being of society and the environment.

Foresight on the future of the Petri dish and microbiology

In looking ahead to the future of the Petri dish and microbiology, extraordinary advances are on the horizon that could radically transform our understanding and application of microbiology. With continued progress in technology and scientific knowledge, a fascinating and promising outlook is anticipated:

Bioprinting of microorganisms: Bioprinting of microorganisms in Petri dishes is expected to enable the creation of three-dimensional microbial structures with unprecedented precision. This would revolutionise fields such as synthetic biology and tissue engineering, opening up new possibilities in biofuel production, drug manufacturing and tissue regeneration.

Microbial Nanotechnology: The integration of nanotechnology into Petri dishes could lead to ultra-sensitive microbial sensors capable of detecting and responding to minute environmental changes. These sensors could have applications in early disease detection, environmental monitoring and food quality control.Microbial Computing: The development of computing systems based on micro-organisms grown in Petri dishes is envisaged. These systems could harness the ability of micro-organisms to process information and perform calculations efficiently, opening up new avenues in biological computing and distributed intelligence. Advanced Microbial Synthesis: Genetic engineering and DNA synthesis are expected to enable the creation of micro-organisms specifically designed to perform complex tasks, such as synthesising chemical compounds or degrading environmental pollutants. These advances could revolutionise industry, agriculture and environmental biotechnology.Microbial Space Exploration: With the rise of space exploration, it is anticipated that Petri dishes will be used as fundamental tools to investigate the presence and activity of micro-organisms on other planets and celestial bodies. This could provide crucial information about the origin of life and the possibilities for habitability in the universe.Personalised Microbial Therapy: Advances in genomic sequencing and data analysis are expected to lead to the development of personalised microbial therapies based on each individual's microbiome. These therapies could be used to treat a wide range of diseases. diseases ranging from metabolic disorders to autoimmune diseases and cancer. Predictive Microbial Simulation: The integration of advanced computational models with experimental petri dish data could enable accurate and predictive simulation of complex microbial communities. This would facilitate the understanding of microbial dynamics in diverse environments and the design of more effective therapeutic and environmental interventions.Artificial Intelligence for Microbial Analysis: Artificial intelligence algorithms trained on large microbiological datasets are expected to significantly improve the ability to analyse and predict microbial phenomena in Petri dishes. This would accelerate drug discovery, microbiome engineering and the understanding of microbial biodiversity.Microbial culture microchips: Microbial culture microchips, which integrate multiple culture chambers on a silicon substrate, are expected to replace conventional Petri dishes. These devices will allow parallel analysis of thousands of microbial samples with greater precision and efficiency.3D culture systems: Three-dimensional culture systems, which better mimic natural cellular architecture, will allow the cultivation of micro-organisms in more realistic environments, which will facilitate research into complex microbial interactions and the development of advanced regenerative therapies. Microbial bioprinting: Bioprinting technology will enable three-dimensional printing of complex microbial structures, opening up new possibilities in microbial tissue engineering and the production of customised bioproducts.

Integrated sensors: Petri dishes of the future are expected to be equipped with integrated sensors that continuously monitor key variables such as pH, temperature, oxygen and nutrient concentrations, providing real-time data on microbial growth and activity.

Integrated omics analysis: The integration of multiple omics techniques, such as genomics, proteomics, metabolomics and metagenomics, in the analysis of microbial samples will allow a more complete understanding of microbial diversity and function in diverse environments.AI-based predictive modelling: Predictive modelling based on artificial intelligence and machine learning will enable accurate prediction of microbial community dynamics in response to environmental changes, facilitating the engineering of customised microbiomes and the optimisation of biotechnological processes.

Applied nanotechnology: Nanotechnology applied to Petri dishes will enable the construction of functional nanomaterials that interact specifically with micro-organisms, facilitating the detection, diagnosis and treatment of infectious diseases.

Quantum cryptography for data security: In the near future, quantum cryptography will be applied to microbiological data security to ensure the integrity and confidentiality of information generated from Petri dish experiments, especially in the context of biosafety and biosecurity.

Programmable microbial nanobots: Programmable microbial nanobots capable of manipulating individual cells in Petri dishes will be developed. These nanobots will be able to perform specific tasks, such as targeted drug delivery or precise genetic modification of micro-organisms. Miniature microfluidic bioreactors: Miniature microfluidic bioreactors integrated into Petri dishes will allow the simulation of complex microenvironments and real-time monitoring of microbial interactions on an unprecedented scale.Augmented reality for microbiological visualisation: Augmented reality systems will be developed that enable researchers to visualise and manipulate microbial samples in Petri dishes in a three-dimensional and collaborative way, improving understanding and scientific communication.

Real-time gene editing: Real-time gene editing will be implemented inside Petri dishes, allowing precise modification of microbial genomes while observing their response in real time, revolutionising genetic engineering and synthetic biology.

Cultivation of extraterrestrial micro-organisms: Petri dishes will be adapted for the cultivation of micro-organisms in simulated extraterrestrial environments, enabling research into space microbiology and the search for life on other planets.

Bioprinting of microbial organs: Using advanced bioprinting techniques, it will be possible to print whole microbial organs in Petri dishes, which could be used to study the interaction between micro-organisms and human tissues in a controlled environment.

Brain-microbe interface: Brain-microbe interfaces will be developed to enable two-way communication between the human brain and microbial communities in Petri dishes, which could have applications in neuroscience and regenerative medicine. Quantum teleportation of samples: Quantum teleportation technologies will be explored to transfer microbial samples instantaneously between Petri dishes located in different parts of the world, facilitating scientific collaboration on a global scale.These advances represent a futuristic vision of how the Petri dish and microbiology could evolve in the distant future, pushing scientific research into new frontiers of discovery and understanding of the microbial world.

BIBLIOGRAPHICAL REFERENCES

1. Petri JR. Eine Kleine Modification Des Koch'Schen Plattenverfahrens (English Translation, Braus, 2020) (Internet). 1887 (cited 21 March 2024). Available from: http://archive.org/details/1887-petri-eine-kleine- modification-des-koch-schen-plattenverfahrens-2020-braus-

2. Rish. The Biomedical Scientist. 2017 (cited 22 March 2024). The big story: the petri dish. Available at: https://thebiomedicalscientist.net/science/big-story-petri-dish

3. Cantón R, Loza E, Romero J. (Applicability of new diagnostic techniques in microbiology; technological innovation). Rev Espanola Quimioter Publicacion Of Soc Espanola Quimioter. September 2015;28 Suppl 1:5-7.

4. Shama G. The "Petri" Dish: A Case of Simultaneous Invention in Bacteriology. Endeavour. March 2019;43(1-2):11-6.

5. Sánchez-Lera RM, Pérez-Vázquez IA. Pasteur and Koch: the fathers of microbiology. 16 Apr. 2022;61(283):1-7.

6. Mahajan M. Etymology: Petri Dish. Emerg Infect Dis. Jan 2021;27(1):261.

7. da Silva JAT. The Misrepresentation of Petri Dish, as "petri" Dish, in the Scientific Literature. Stud Hist Sci. 2023;(22):611-26.

8. Julius Richard Petri (1852-1921) (Internet). (cited 22 March 2024). Available from: https://www.historiadelamedicina.org/petri.html

9. Sauka DH. Julius Richard Petri: The creator of the plates we use every day. April 2011 (cited 22 March 2024); Available from: https://ri.conicet.gov.ar/handle/11336/192126

10. Koch R. Zur Untersuchung von pathogenen Organismen. Norddeutschen Buchdruckerei und Verlagsanstalt; 1881. 82 p.

11. Cornil V, Babeș V. Les bactéries et leur role dans l'anatomie et l'histologie pathologiques des maladies infectieuses. Germer Bailliere et cie.; 1886. 964 p.

12. Worboys M. Robert Koch: a life in medicine and bacteriology. Med Hist. July 1990;34(3):347-8.

13. Dehnhardt WL. A personal history of bacteria (Internet). RIL editores; 2007 (cited 12 April 2024).

14. Friedman M, Friedland GW. Medicine's 10 greatest discoveries (Internet). Yale University Press; 1998 (cited 12 April 2024).

15. Furones MD. Sampling for antimicrobial sensitivity testing: a practical consideration. Aquaculture. May 15, 2001;196(3):303-9.

16. van Belkum A, Burnham CAD, Rossen JWA, Mallard F, Rochas O, Dunne WM. Innovative and rapid antimicrobial susceptibility testing systems. Nat Rev Microbiol. May 2020;18(5):299-311.

17. Atlas RM. Handbook of media for environmental microbiology (Internet). CRC press; 2005 (cited 12 April 2024). Available from: https://www.taylorfrancis.com/books/mono/10.1201/9781420037487/ha ndbook-media-environmental-microbiology-ronald-atlas.

18. Adams MR, Moss MO. Food microbiology (Internet). Royal society of chemistry; 2000 (cited 12 April 2024).

19. Gilmore BF, Denyer SP. Hugo and Russell's pharmaceutical microbiology (Internet). John Wiley & Sons; 2023 (cited 12 April 2024).

20. Grote M. Petri dish versus Winogradsky column: a longue durée perspective on purity and diversity in microbiology, 1880s-1980s. Hist Philos Life Sci. Nov 29, 2017;40(1):11.

21. History of the Petri dish - PHOENIX BIOMEDICAL PRODUCTS (Internet). (cited 14 April 2024). Available from: https://phoenix-

biomed.com/history-of-the-petri-dish/

22. Mangiarotti A, Caretta G, Nelli E, Piontelli E. Biodeterioration of plastic materials by microfungi. Bol Micologico. January 1, 1994;9:39-47.

23. Sneath PHA, Stevens M. A Divided Petri Dish for Use with Multipoint Inoculators. J Appl Bacteriol. 1967;30(3):495-7.

24. Ingham CJ, van den Ende M, Pijnenburg D, Wever PC, Schneeberger PM. Growth and multiplexed analysis of microorganisms on a subdivided, highly porous, inorganic chip manufactured from anopore. Appl Environ Microbiol. Dec 2005;71(12):8978-81.

25. Cooper WG. A Modified Plastic Petri Dish for Cell and Tissue Cultures. Proc Soc Exp Biol Med. April 1, 1961;106(4):801-3.

26. Bolafi A, Litsky W. STUDIES ON THE USE OF PLASTIC PETRI DISHES J Food Prot. 1 March 1959;22(3):67-70.

27. Reeder JC, Shakespeare AP, Bryan C, Keaney MG, Ganguli LA. Comparison of three-compartment Petri dishes and individual plates for routine culture of vaginal swabs. J Clin Pathol. Nov 1990;43(11):947-9.

28. Sridhar A, de Boer HL, van den Berg A, Le Gac S. Microstamped Petri dishes for scanning electrochemical microscopy analysis of arrays of microtissues. PloS One. 2014;9(4):e93618.

29. Lovelace TE, Colwell RR. A multipoint inoculator for petri dishes. Appl Microbiol. June 1968;16(6):944-5.

30. Tiam Kapen P, Fotsing Kwetche PR, Youssoufa M, Kayo Mbomda WC, Ketchogue RM, Ganwo Dongmo S. An automatic multipoint inoculator for the determination of minimum inhibitory concentrations (MICs) of antibiotics in low-income countries: a technical note. Australas Phys Eng Sci Med. Dec 2019;42(4):905-12.

31. Uber DC, Jaklevic JM, Theil EH, Lishanskaya A, McNeely MR. Application of robotics and image processing to automated colony picking and arraying. BioTechniques. November 1991;11(5):642-7.

32. Cooper JM. Towards electronic Petri dishes and picolitre-scale single-cell technologies. Trends Biotechnol. June 1999;17(6):226-30.

33. Antonios K, Croxatto A, Culbreath K. Current State of Laboratory Automation in Clinical Microbiology Laboratory. Clin Chem. Jan 1, 2022;68(1):99-114.

34. Burckhardt I. Laboratory Automation in Clinical Microbiology. Bioengineering. December 2018;5(4):102.

35. Croxatto A, Prod'hom G, Faverjon F, Rochais Y, Greub G. Laboratory automation in clinical bacteriology: what system to choose? Clin Microbiol Infect. Mar 1, 2016;22(3):217-35.

36. Bourbeau PP, Ledeboer NA. Automation in Clinical Microbiology. J Clin Microbiol. Dec 21, 2020;51(6):1658-65.

37. Novak SM, Marlowe EM. Automation in the clinical microbiology laboratory. Clin Lab Med. September 2013;33(3):567-88.

38. Jacot D, Sarton-Lohéac G, Coste AT, Bertelli C, Greub G, Prod'hom G, et al. Performance evaluation of the Becton Dickinson KiestraTM IdentifA/SusceptA. Clin Microbiol Infect. Aug 1, 2021;27(8):1167.e9-1167.e17.

39. Gao J, Chen Q, Peng Y, Jiang N, Shi Y, Ying C. Copan Walk Away Specimen Processor (WASP) Automated System for Pathogen Detection in Female Reproductive Tract Specimens. Front Cell Infect Microbiol (Internet). November 17, 2021 (cited April 7, 2024);11. Available from: https://www.frontiersin.org/articles/10.3389/fcimb.2021.770367

40. Croxatto A, Dijkstra K, Prod'hom G, Greub G. Comparison of Inoculation with the InoqulA and WASP Automated Systems

with Manual Inoculation. J Clin Microbiol. July 2015;53(7):2298-307.

41. Baker J, Timm K, Faron M, Ledeboer N, Culbreath K. Digital Image Analysis for the Detection of Group B Streptococcus from ChromID Strepto B Medium Using PhenoMatrix Algorithms. J Clin Microbiol. Dec 17, 2020;59(1):e01902-19.

42. Jones D, Cundell T. Method Verification Requirements for an Advanced Imaging System for Microbial Plate Count Enumeration. PDA J Pharm Sci Technol. Mar 1, 2018;72(2):199-212.

43. Strauss S, Bourbeau PP. Impact of introduction of the BD Kiestra InoqulA on urine culture results in a hospital clinical microbiology laboratory. J Clin Microbiol. May 2015;53(5):1736-40.

44. Graham M, Tilson L, Streitberg R, Hamblin J, Korman TM. Improved standardization and potential for shortened time to results with BD KiestraTM total laboratory automation of early urine cultures: A prospective comparison with manual processing. Diagn Microbiol Infect Dis. September 2016;86(1):1-4.

45. Cherkaoui A, Renzi G, Martischang R, Harbarth S, Vuilleumier N, Schrenzel J. Impact of Total Laboratory Automation on Turnaround Times for Urine Cultures and Screening Specimens for MRSA, ESBL, and VRE Carriage: Retrospective Comparison With Manual Workflow. Front Cell Infect Microbiol. 2020;10:552122.

46. Kempner ME, Felder RA. A Review of Cell Culture Automation. JALA J Assoc Lab Autom. April 1, 2002;7(2):56-62.

47. Lin E, Liu E, Pan M, Reyes K. Automated Petri Dish Filler Machine. April 2023 (cited April 14, 2024); Available from: http://deepblue.lib.umich.edu/handle/2027.42/177464

48. Marotz J, Lübbert C, Eisenbeiß W. Effective object

recognition for automated counting of colonies in Petri dishes (automated colony counting)1. Comput Methods Programs Biomed. September 1, 2001;66(2):183-98.

49. Faiña A, Nejati B, Stoy K. EvoBot: An Open-Source, Modular, Liquid Handling Robot for Scientific Experiments. Appl Sci. Jan 2020;10(3):814.

50. Wu Z, Xu Q, Ai N, Ge W. Design of a Novel Magnetically Actuated Biaxial Robot With Compact Structure and Easy Operation. IEEE Robot Autom Lett. June 2023;8(6):3884-91.

51. Lange O, Erhard M, Teutsch C, Sander J. MIROB: automatic rapid identification of micro-organisms in high through-put. Troccaz J, editor. Ind Robot Int J. January 1, 2008;35(4):311-5.

52. Naugler C, Church DL. Automation and artificial intelligence in the clinical laboratory. Crit Rev Clin Lab Sci. Feb 17, 2019;56(2):98- 110.

53. Smith KP, Kirby JE. Image analysis and artificial intelligence in infectious disease diagnostics. Clin Microbiol Infect. Oct 1, 2020;26(10):1318-23.

54. Ingham CJ, Sprenkels A, Bomer J, Molenaar D, van den Berg A, van Hylckama Vlieg JET, et al. The micro-Petri dish, a million-well growth chip for the culture and high-throughput screening of microorganisms. Proc Natl Acad Sci U S A. November 13, 2007;104(46):18217-22.

55. Forbes BA. Microbiological diagnosis (Internet). Ed. Médica Panamericana; 2009 (cited 12 April 2024).

56. Tian Z, Wang Z, Zhang P, Naquin TD, Mai J, Wu Y, et al. Generating multifunctional acoustic tweezers in Petri dishes for contactless, precise manipulation of bioparticles. Sci Adv. Sep 11, 2020;6(37):eabb0494.

57. Lahoz-Beltrá R. Bioinformatics: Simulation, artificial life and artificial intelligence (Internet). Ediciones Díaz de Santos; 2004 (cited 12 April 2024).

58. Hernández M, Quijada NM, Rodríguez-Lázaro D, Eiros JM. Application of massive sequencing and bioinformatics to clinical microbiological diagnosis. Rev Argent Microbiol. 2020;52(2):150-61.

yes I want morebooks!

Buy your books fast and straightforward online - at one of world's fastest growing online book stores! Environmentally sound due to Print-on-Demand technologies.

Buy your books online at
www.morebooks.shop

Kaufen Sie Ihre Bücher schnell und unkompliziert online – auf einer der am schnellsten wachsenden Buchhandelsplattformen weltweit! Dank Print-On-Demand umwelt- und ressourcenschonend produzi ert.

Bücher schneller online kaufen
www.morebooks.shop

info@omniscriptum.com
www.omniscriptum.com